THE

CLARA NEVADA

THE CLARA NEVADA

Gold, Greed, Murder and Alaska's Inside Passage

STEVEN C. LEVI

THE History PRESS

Published by The History Press
Charleston, SC 29403
www.historypress.net

Front cover image: While the water in the Lynn Canal can be calm, as seen here, when there is a storm, twelve-foot waves and wind exceeding ninety miles per hour are not uncommon. Photo by Peter Metcalfe. *Courtesy of the State of Alaksa Department of Commerce and Economic Development.*

First published 2011
Manufactured in the United States
ISBN 978.1.60949.288.5

Library of Congress Cataloging-in-Publication Data

Levi, Steven C.
The Clara Nevada : gold, greed, murder, and Alaska's Inside Passage / Steven C. Levi.
p. cm.
Includes bibliographical references and index.
ISBN 978-1-60949-288-5
1. Clara Nevada (Steamboat)--Explosion, 1898. 2. Explosions--Alaska--History--19th century. 3. Shipwrecks--Alaska--History--19th century. 4. Shipping--Washington (State)--Seattle--History--19th century. 5. Avarice--Social aspects--Case studies. 6. Seattle (Wash.)--History--19th century. 7. Lynn Canal (Alaska)--History--19th century. 8. Alaska--History--1867-1959. 9. Inside Passage--History--19th century. 10. Klondike River Valley (Yukon)--Gold discoveries. I. Title.
F909.L434 2011
979.8'02--dc22
2011010590

Dedicated to the passengers and crew of the Clara Nevada *who may well have been the victims of one of the largest robberies in American history.*

Contents

1. The Maritime Rush to Skagway and Dyea 9
2. The Wreck of the *Clara Nevada* 17
3. The *Seattle Times* and *Post-Intelligencer* Do Battle 27
4. "Curiouser and Curiouser" 35
5. The Search for Answers 39
6. What Happened to C.H. Lewis? 47
7. The End of the Maritime Frontier 65
8. Mass Murder and Robbery or an Accident? 73

Appendix 93
Notes 95
Bibliography 103
About the Author 107

Chapter 1

The Maritime Rush to Skagway and Dyea

The saga of the *Clara Nevada* actually began in Canada two years before the ship was even launched. In May 1896, prospector George Carmack and his brothers-in-law, Skookum Jim and Tagish Charley, stumbled onto one of the richest gold finds in North American history at Rabbit Creek, one of the numerous small streams leading into the remote Thron-diuck River. Skookum Jim made the discovery while cleaning a dish pan in the waters of the creek. That dish pan yielded four dollars. The next day, August 17, 1896, the three men staked their claim.

The news of the strike spread quickly, and soon the creek banks were packed with men digging for the yellow metal. Rabbit Creek was triumphantly renamed Bonanza Creek, and Dawson City sprang to life at the confluence of the now-renamed Klondike River and mighty Yukon.

About a year later, in July 1897, news of the strike reached the United States in dramatic fashion: Tom Lippy stepped off the steamboat *Excelsior* in San Francisco with $50,000 in gold dust. Little did he know that he was going to change the face of the age—he set off a clap of thunder heard 'round the world.

If there was any doubt as to the existence of the Klondike Strike, it was extinguished two days later when the *Portland* docked in Seattle carrying more gold dust–laden prospectors. Though it was only carrying $700,000 in gold, according to the *Seattle Post-Intelligencer*, the *Chicago Tribune* increased the quantity to "a ton." There was so much gold on board, a Seattle paper noted, that some gold had been smuggled aboard the *Portland* to avoid the freight charge. The *San Francisco Call* declared, "Nuggets are reported as big as potatoes." The story was flashed across the country and around the world. Within two weeks, more than 1,500 prospectors had left Seattle for the Klondike, and there were nine more ships in the harbor ready to sail. The greatest gold rush in North American history was on!

Within days of the arrival of the *Portland*, Seattle streets began teeming with would-be millionaires seeking passage north. Thousands of men—and more than a few women—left their jobs without so much as a day's notice. Thousands grabbed their valises and bedrolls and headed to Portland and Seattle, the gateways to the Klondike.

But the era of the Klondike Strike was short, barely sixteen months. What began with the arrival of the *Portland* on July 17, 1897, ended in May 1899 with the news of the gold strike in Nome. Stampeders left Dawson by the thousands for the golden beaches of Nome.

From the four corners of the earth, Argonauts—the name itself coined from the ship of the legendary Greek hero Jason in quest of the golden fleece, the *ARGO*—converged on Seattle, Portland and Vancouver with a single thought in mind: passage north. All that mattered was getting to the gold fields as fast as possible. Each day was a fortune lost, a day in which someone else might find the Eldorado.

Getting to the Klondike, however, was a bit more difficult than just getting to the Pacific Northwest and catching a ship. The fastest way was by ship, but not everyone could afford passage. Those with less money used whatever means they could afford. Some came by foot, others by bicycles or canoe. Unusual modes of transportation included rowboat, balloon, airplane and even reindeer.[1] No means of transportation was considered too bizarre to be considered.

By far the most popular route was by ship, from Seattle to the twin Alaska boomtowns of Dyea and Skagway. Here the stampeders disembarked and made their way up the infamous Chilkoot Pass ice stairway or the longer but less steep White Pass. Once the passes had been crested and the Argonauts were inside Canada, they would follow the rivers and chain of lakes northward to Dawson, the heart of the gold fields.

The gold route over the Chilkoot and White Passes is best known today courtesy of a band of outlaws who ruled Skagway. A gang led by the nefarious Jefferson Randolph "Soapy" Smith found that it was infinitely easier to fleece stampeders than to expose themselves to the hard labor of digging for nuggets. Those whom the gang could not roll, they robbed; those they could not rob, they hoodwinked. Soapy Smith and his gang ruled Skagway, despoiling and hoodwinking stampeders with impunity until July

8, 1898, the day Soapy was shot dead on the Skagway dock. Leaderless, the gang scattered and lost its menace.

But the trip to Skagway was often more hellish than dealing with Soapy Smith and his minions. The *Clara Nevada* was the epitome of what could go wrong on the Inside Passage, the shortest sea route from Seattle to Skagway and Dyea, the salt-water gallery of channels protected from the turbulent storms of the open Pacific.

During the summer, the waters were calm and storms more an annoyance than a danger. To this day, summer cruises up the Inside Passage are one of the most popular summer luxury liner routes. Whales, majestic scenery and quaint fishing villages charm visitors to Alaska, while the weather is mild enough to allow jogging on the open top decks of the pleasure ships.

But during the winter, the Inside Passage can be exceedingly treacherous. Funneled by the islands, ferocious Pacific winds blast through the straits and barrel up the channels. Known as *takus*, such winds can easily roll a poorly loaded ship. Frigid winds can coat a ship with ice in a few hours. Snow can fall so heavily that whiteout conditions exist, and fog can hang so low that the rugged shoals of the channels cannot be seen until a luckless ship runs aground.

In the Klondike stampede of 1898, as in all other gold rushes, entrepreneurs learned quickly that there was more money to be had from the pockets of the stampeders than out of ground. The greatest scams were not in the sales of fry pans, trousers, flour or shovels; they were in transportation. Stampeders had to go north—a high demand; the number of ships was limited—a low supply. Thus, by that natural mechanism, the shipping industry was destined to make money—lots of money.

Seattle and Portland also had their Soapy Smiths—too many of them. In the mad rush to provide transportation, every ship on the Pacific coast that could be found, regardless of its seaworthiness, was launched into the trade. Here was a fleet born of greed. Rotting derelicts were yanked off beaches or unearthed from maritime graveyards. Refitted, they became members of a doom-haunted flotilla. Sailing vessels, blockade runners, private yachts and decrepit steamers of every description were swiftly refurbished, repainted and rechristened to be offered to the gold-dazzled public. Often these repairs were slight.

Many of these vessels could hardly be called safe even when upgraded. They were loaded for profit, not safety. Holds were crammed dangerously full. Hallways were lined with swaying, unsteady crates. Deck cargo was piled as high as ropes could reach. Uneven loading often caused ships to list dangerously. Sometimes even the hatches were left open and crammed with cargo. Should the ship encounter a storm, the open hatches could spell doom.

Adding to the mélange, part of the cargo was often animal. On some ships, horses, dogs and cattle were taken onboard first, and human passengers were squeezed into the remaining space. Often this was not much. The epitome of overloading was the *Al-Ki*, which left Seattle on July 19, 1898. The ship carried 110 human passengers, 900 sheep, 65 cattle, 30 horses and still managed to load 350 tons of supplies. Another dilapidated wreck, the *Colorado*, left Seattle with 350 horses, 150 cattle and 100 dogs. Securing animals and cargo was frequently done without much care, even on the larger ships. When the *Queen* went aground in Wrangell Narrows, the ship tilted to forty degrees. Falling baggage crushed a number of horses and injured many more. A passenger on the *Islander*, stampeder Alexander Whyte, wrote to his wife that the steamer was so packed it had a "100 ton overload," including "about 600 dogs and 8 oxen and 50 to 75 horses," besides the "420 to 450 passengers and crew." Whyte estimated that was about 150 passengers too many, but the steamer company didn't seem to care because the transportation business was just "a skin game all through."

Humans were often treated as just another form of troublesome cargo. If staterooms were full, passengers had to find sleeping quarters between crates, in the engine room or wherever else a human body could wriggle. The Canadian steamer *Amur* overbooked its passengers by a factor of five. In first class, rooms that normally held two were booked to ten. Second class was substantially more crowded. Passengers had to seize any open bed if they wanted to sleep. To eat, the stampeders had to snatch food from trays as they passed down the hallway on their way to the dining room, a salon that could only accommodate twenty-seven people at a time.

In Dyea, customs officers reported "nearly every vessel arriving here carr[ied] twice the passengers the law allow[ed] it to carry and many of them are condemned craft that have been fitted up for this trade."[2] Tales of raw terror onboard were commonplace. One passenger described his voyage north as "a floating bedlam, Pandemonium let loose, the Black Hole of Calcutta in an Arctic setting."

Second-class quarters were likened to "a veritable beehive [where] the atmosphere resembled that of a dungeon in Afghanistan." Fred Kimball, postmaster of St. Michael during the Klondike Rush, reported seeing the *Homer* at the Dutch Harbor dock "loaded with freight so that not a ray of daylight could get into a stateroom." Had the ship gone down, "her passengers would have stood no more a show than rats in a trap."

Passengers could be a danger, too. Many were not the least concerned with overloading. They wanted to go north immediately. This was a gold rush. Every hour spent on a Seattle dock meant that someone else was staking the best claims. Passengers often proved cantankerous, even if their demands meant placing the ship in danger. On August 18, 1897, as an example, the *Post-Intelligencer* reported a full-scale passenger riot when the *Humboldt* attempted to leave with the passengers but not their supplies. Threats of "gun play, mutiny and violence" kept the ship from sailing. The cargo was finally loaded.

Rebellions and mutinies were not unknown. Petitions filled out by passengers and given to a government official at a port were not unusual. But these were marginally effective. All in all, once passengers were on board, they were at the mercy of the hard-bitten captains and crews. One surviving petition concerns the overselling of first-class cabin space on the steamer *Protection*. The ship was licensed for sixteen first-class passengers, but the company had sold that space to twenty-seven people. There is no evidence that the United States government officials in St. Michael in August 1898 did anything other than accept the petition. The only proof that they did even that is due to the document still being in existence.

Passengers were not above threats of violence themselves. Even unarmed, they could be violent. In September 1899, Captain O'Brien of the *Bark Hunter* arrived in Seattle, making "unusually quick time." The passengers, angry over the quality of the food on board, had "talked of hanging the master to the yardarm of the vessel."

Then there were the crews—or what were called "crews." When hundreds of seasoned sailors abandoned deck timbers for the gold fields, their places had to be filled by men whose maritime experience was limited to rowboats on calm country lakes. Unscrupulous men claimed years of sea experience to get a free ride north. Farmers, bartenders, card sharks and derelicts signed on as crew, many of them so unfamiliar with the sea they were unsure what their job title entailed.

Officers were not much better. Crew members were beaten, brutalized and punished unmercifully. Men were beaten to the point where they were permanently crippled, and few captains were punished for the crimes. The captain of the schooner *Lyman D. Foster*, as an example, was the subject of a libel suit that "read much like a page from a mildly colored dime novel." The captain was accused of "inveig[ing]" three seamen onto his ship and then "shanghai[ng]" them. Some captains were out-and-out cowards. When the schooner *Elsie* was about to wreck on Chichagof Island, the captain rushed out on deck without hat or shoes "wild with fear. He knelt down on the icy deck and with tears streaming down his face cried out 'God Help Us!'"[3]

Often captains were as bad as their crew. Take the case of the *Blakley*, a square-rigged brigantine. The ship had been rotting on the beach in Tacoma for two years before it was snatched up by four Klondike exploration syndicates. The years on the beach had done the vessel permanent damage, and even when refitted, it was in such poor shape that the captain "refuse[ed] to take charge until he had consumed the [ship's] entire supply of whiskey."

But officers and crew members were desperately needed, regardless of their inexperience. In March 1898, forty-two steamers and twenty-two sailing vessels and barges were made available for the Seattle to Dyea and Skagway run. The round trip took about fifteen days, and the carrying capacity of those ships was 12,472 passengers and 53,570 tons of cargo. This meant that there could have been as many as 20,000 passengers heading north that month. Considering that the ships were overloaded and overbooked, the actual number may have been substantially higher. Each of those vessels needed a crew. With so much money to be made, captains could not be choosy about whom they hired.

Business in Seattle boomed. But there was a downside to the sudden surge of economic activity in Seattle. While the Seattle Chamber of Commerce may have been pleased with the influx of business, many mariners were quite concerned with the impact greed was having on their industry. Ships that should have been allowed to rot gracefully to splinters were being pressed into service in waters that were among the world's most treacherous. Captains of dubious ability were commanding crews of questionable competence in ships so decrepit that the first storm could easily reduce them to timbers. Many were so reduced.

These mariners became even more concerned when editorials in East Coast papers—the home ports of their competitors—began stating the obvious. In February 1898, the Providence, Rhode Island *Journal* touched a sensitive nerve when it stated that "in the rush to gold fields any old thing, so long as it would float, is sent forth." In response to this and other articles and editorials, Erastus Brainerd, head of the publicity committee for the Seattle Chamber of Commerce, wrote the supervising inspector of the Steamboat Inspection Service in Washington D.C., demanding to know what the service was going to be doing in Puget Sound that summer and whether the inspections of steamboats "differ from or [are] inferior to the inspection of steamers in other parts of the United States."

James Dument, the supervising inspector general in Washington D.C., replied that "the fact that the inspectors [in Seattle] still [are] retained should be prima facie sufficient evidence of the opinion of this office as to such general charges." The *Post-Intelligencer* ran both letters, one after the other, at the bottom of a lengthy piece on the *Clara Nevada*, one of the first ships to go to the bottom in 1898.[4]

Such attempts at caution were ill-placed at a time where there was so much easy cash to be made. Pure greed was driving the gold rush juggernaut. Arthur Dietz from New York wrote that the "Devil reigned supreme" in Seattle. It was a town gone mad with "Klondicitis." This, then, was the setting for the tragedy that claimed the *Clara Nevada*.

Chapter 2

The Wreck of the *Clara Nevada*

On the evening of February 5, 1898, the night the *Clara Nevada* left Skagway, every possible maritime factor was mitigating against the ship. It was an old, poorly refitted and marginally operational vessel. Its crew was of questionable ability and virtue. It was steaming the turbulent winter waters of Alaska's Inside Passage, running before a ninety-mile-an-hour wind churning up waves cresting at twelve to fifteen feet. It was also snowing. Such a night, mariners today remark, would have made a "dead fisherman quiver."

For the first ten miles, the vessel navigated Taiya Inlet, a narrow, glacier-formed passage that ran a shade west of due south. Then she entered the waters of Lutak Inlet, carefully avoiding the shoals at Indian Rock, the only midchannel hazard in the Chilkoot Inlet that runs just a shade east of due south. On the port side—left and, in this case, east—ran the mainland. On the other side was the long arm of the Chilkat Peninsula, which ended at Seduction Point, so named for the "designing nature of the Indians there." The geomorphic extension of land thrusting out into the Chilkoot Inlet from Seduction Point was known as "Seduction Tongue," leaving historians to ponder whether this was meant as a double entendre.

Once out of the Chilkoot Inlet, the *Clara Nevada* entered the Lynn Canal, another glacier-formed corridor that also ran close to due south for fifty miles before emptying into the confluence of Chatham Strait coming from the south and Icy Strait from the west. Eldred Rock, the final resting place of the *Clara Nevada*, lies midchannel, ten miles south of Seduction Point. To the east of Eldred Rock is the mainland. To the west is Sullivan Island, ironically named for the captain of the *Louisa Downs*, wrecked on the island in 1867.

Hardly noteworthy in life, in death the *Clara Nevada* has become immortal. Built in 1872 as a survey vessel for the United States Coast and Geodetic Survey (USCGS), she was named the *Hassler* in honor of Ferdinand Hassler, the first superintendent of the survey (1807–18 and 1832–43). For twenty-five years, she plied the waters of Alaska as her crew mapped the coastline. In addition to locating and mapping hazards to maritime travel, she was also instrumental in logging many of the names of prominent geographic features along Alaska's lengthy coast. In 1882, her name was even added to the topography of the northland. Surveyor William Healy Dall, at one time the acting assistant of the U.S. Coast Survey, named Hassler Point on the northeast coast of Dall Island.

In 1897, when the USCGS felt that the *Hassler* was too old to be considered seaworthy, she was sold for $15,700 to the Pacific & Alaska Transportation Company. Based out of Portland, this was one of the numerous, rapidly formed maritime companies that were taking advantage of the sudden demand for transportation to the Klondike gold fields.

As with many such firms, the Pacific & Alaska Transportation Company had a less than sterling reputation. Prior to owning the *Clara Nevada*, the company—owned by two McGuire brothers—had been part of another company, the Portland & Alaska Trading & Transportation Company. Under that corporate charter, they had purchased a Columbia River stern-wheeler, the *Eugene*, which they intended to have towed across the open waters of the north Pacific to St. Michael. From there, the shallow draft vessel would have been used to transport Argonauts up the Yukon River to Dawson. The job of towing the *Eugene* had been assigned to Captain McIntyre and the *Bristol*. The captain of the stern-wheeler was C.H. Lewis, later to be the skipper of the *Clara Nevada*.

The saga of the *Eugene* was destined to reflect both comedy and drama. It was agreed that the ships would rendezvous in Union Bay, British Columbia. The stern-wheeler arrived ahead of the *Bristol* and was immediately impounded by British authorities "on various charges." When the *Bristol* arrived, it refused to throw the *Eugene* a tow line until these "customs difficulties" were resolved. This did not please Captain Lewis, who felt Canadian authorities would probably never allow the stern-wheeler out of port—for good reason. Rather than abandon the *Eugene*, Lewis ordered a "full head of steam, slipped his lines" and headed for St. Michael, alone, across the north Pacific with a full complement of passengers.

(The fastest—and most expensive—route to the Klondike was up the Yukon River. Steamers would leave Seattle and head across the north Pacific to Dutch Harbor in the Aleutian Islands of Alaska. The vessels would then head north across the Bering Sea to St. Michael near the mouth of the Yukon River. Usually the arrival at St. Michael meant that passengers were switching from an ocean-going steamer to a river boat. River boats, like the *Eugene*, had shallow drafts and were not stable in the ocean. If waves rose, the ships could be flipped.)

Once the British customs agents realized that Lewis and the *Eugene* had given them the slip, they sent an old steamer, the *Hope*, in pursuit. But the *Hope* was no match for the speedy stern-wheeler and the determined Captain Lewis, who swore he would "burn all the bacon aboard before I let [the British authorities] catch us."

The *Bristol* also set off in pursuit of the *Eugene*, primarily because the *Eugene* was a riverboat that would break apart in a storm on the open sea. Captain McIntyre had good cause for concern: a storm was brewing. The *Bristol*, however, could hardly have been called seaworthy. It had left Victoria with six hundred horses in two-foot-wide stalls. Passengers were sleeping on plank bunks anchored wherever there was room for a man to stretch out and in many places where a man could not. She was so badly overloaded, she almost turned turtle a short distance out of Seattle.

Even though she was heavily loaded, the *Bristol* was able to catch the *Eugene* near Seymour Narrows,[5] where she threw a cable to the stern-wheeler. This cable came none too soon. A heavy southeast blow came up shortly, "springing the stern-wheeler's timbers and causing her to work badly and take in water at an alarming rate."

While Captain Lewis swore he was going on to St. Michael, his passengers felt otherwise. Understandably frightened, they held a mass meeting and, when assembled, demanded that the *Eugene* return to port. Lewis was recalcitrant. He said he was going on to St. Michael, even "if all hell broke loose." It did. Given no other choice, the passengers, mostly miners, took possession of the *Eugene* and then signaled the *Bristol* to take them aboard before the *Eugene* sank.

This was not as easy as simply transferring passengers, however. The captain of the *Bristol* refused to take any passengers on board until they signed "a release of any cause of action which they might have acquired against the *Bristol*." The passengers were reluctant to do so but, with the approaching storm, they felt they had little choice. After a standoff, the passengers relented and almost all signed the releases. One man, E.B. McFarland, the general manager of the Portland & Alaska Trading & Transportation Company, refused to sign. This so enraged the rest of the passengers that they threatened to lynch him. With waves churning violently and the ship in danger of sinking at any moment, few passengers were interested in the niceties of legal phraseology. McFarland was "helped to a determination by a committee of passengers who presented guns at his breast and fired pistols within an inch of his ears." He signed.

After the passengers were loaded onto the *Bristol*, the rapidly filling *Eugene* was hauled to Alert Bay, where British customs agents once again seized the vessel and ordered it towed to Victoria, British Columbia.[6] Though his ship was badly damaged, Captain Lewis cut the tow line to the *Bristol* late that night and, once again, disappeared into the Pacific. When he reappeared, the *Eugene* was in Port Townsend on Puget Sound, safely on the American side of the border.

The dramatics were all in vain. Though the owners swore that the *Eugene* would try for St. Michael again, the vessel never got the chance. The fiasco cost the Portland & Alaska Trading & Transportation Company too much money. The company declared bankruptcy, and the *Eugene* had to be returned to her previous owners. Lewis was sued for his actions and McFarland sued for duress.[7]

The bankrupt owners of the *Eugene*, the McGuire Brothers, then purchased the *Hassler* and hired Lewis to captain their newest flagship. The vessel was refurbished and then renamed the *Clara Nevada* after a "well-known Western actress of that period. "

As a historical aside, research has failed to locate any person—real or fictional—named Clara Nevada. The only Nevada that appeared to fit the naming of the ship was Emma Nevada, known as "the world's greatest coloratura singer," a woman alleged to have had an affair with Wild Billy Metson, also known as "Hell Fire Bill from Gold Hill," who later gained the reputation in Alaska as the law "North of 53."[8]

(It is also interesting to note, though of dubious authenticity, that "old sailors who live[d] in an atmosphere of superstition" believed that the *Clara Nevada* was a cursed ship. Worse, she was "doubly hoo-dooed because she had two names ending in A." This, supposedly, was because there were statistically more ships wrecked whose names ended in the letter "A" than with any other letter.[9] A quick check of ships wrecked or sunk in the Pacific Northwest in 1898, oddly, confirmed this superstition. Of the fifty-six such ships listed in McCurdy as having gone to Davy Jones's Locker, thirteen had names ending in "A." "D" and "N" had seven apiece, and "E" was the third most numerous with five members. This, the author speculates, is more a matter of orthography than any insinuation by a supernatural power.)

A fairly large ship considering the vessels plying the waters between Seattle and Skagway but small compared to the *Queen* and other coast liners, the *Clara Nevada* could carry a crew and two hundred passengers, many in first-class accommodations, along with three hundred tons of cargo. Steam powered, she had a length of 154.0 feet, beam of 24.7 and draft of 11.8.[10] It was not a stern- or side-wheeler, so travel across the open ocean was not treacherous, a fact that undoubtedly must have made Captain Lewis breathe easier.

The ship had three masts used for supporting sails. These were, in essence, a back-up power system for the ship. If the engine failed or the ship ran out of coal, the sails could be unfurled.

On her maiden voyage, the *Clara Nevada* left Seattle in a busy week. Over the previous seven days, eight steamers carrying upward of two thousand stampeders had left for Skagway and Dyea, the twin ports at the upper end of Taiya Inlet. But the voyage north was not without incident. As she backed out of her berth in Seattle, the ship collided with the U.S. Revenue cutter *Grant*. Upon her arrival in Port Townsend the next day, she rammed the dock and damaged her bowsprit. She fought a ferocious storm as she headed north and landed in Juneau too late to offload her cargo of dynamite.

At Skagway, according to one witness, J.W. Lansing, M.D., she "jammed against the pier and shattered three or four piles badly; after which [she] backed away and made several other efforts to land, but it kept us about two hours to make a satisfactory landing."[11] She offloaded passengers and cargo in Dyea and Skagway and then steamed south, now running with the wind she had been fighting the day before, headed for Seattle.

What happened next is pure conjecture. A witness on shore at Seward City, now known as Comet, eight miles away from the *Clara Nevada*'s final resting place, reported he saw a ship on fire near Eldred Rock. An orange fireball had sprouted on the water, and "embers were seen to fly into the sky at a great height."[12]

In the next instant, the night was pitch-black. He assumed that a ship had exploded and gone to the bottom of the channel. Quickly, he passed the word but there was nothing anyone could do that night. The storm was near hurricane force, and it would have been impossible for any craft to reach the site of the fireball.

Word was passed to Juneau by the steamer *Coleman*. The deputy collector of the United States Customs in Alaska, Inspector T.A. Marquam, immediately chartered the steamer *Rustler* to investigate. At Comet, he interviewed witnesses and examined debris found along the beach. He reported finding clothing, photographs, furniture and parts of broken boats, along with the following letters, constituting the name of the boat, "——ada."[13]

Other debris included skylights, parts of a deck house and "a lot of 16 foot planks," which the *Clara Nevada* had loaded on her deck.[14] Also found were life preservers and a hatch cover with the name *Hassler* printed on them. This led to some confusion. Since the McGuires stated that the name *Hassler* had been eliminated from all locations on the ship and new life preservers taken onboard, there was speculation that the *Clara Nevada* was not the boat that had gone down. This suggestion was given more substance when it was reported that a passenger on another steamer had spotted the *Clara Nevada* "lying to under the shelter of Douglas Island" on the night the vessel was supposed to have gone down. This sighting clearly proved to have been in error.[15]

When Marquam came to Eldred Rock, he discovered the corpse of a steamer. He reported the "hull was lying in about four fathoms [24 feet] of water, and her outline could be seen distinctly." But, as only the spars could be seen above the water at low tide, there was no way to positively identify the wreck.

A diver sent overboard reported a "great black hole in the hull where the boiler room had been."[16] Marquam then proceeded to "coast around Lynn Canal" looking for survivors. He "examined Chilkat Island thoroughly," as well as "Sullivan Island and the shore of both sides of Lynn Canal, Henry Bay and Shelter Island," but was "unable to find any trace of any one." But he did find the footprints of "some animal in the snow, and as there are no wolves about," he believed that these were the paw prints "of a dog that had swam from the wreck."

Even if there were any corpses to be found, the *Rustler* would have been unable to pinpoint the location of the cadavers. The storm that had just passed had dumped "18 inches of snow" on the area. This would have covered any human remains for days, if not weeks.

Perhaps the most persuasive proof that the vessel that had gone down was the *Clara Nevada* was that the vessel never again made port. Laundry left by the crew in Juneau was never picked up nor was the dynamite destined for the Treadwell Mine in Douglas ever delivered. Combining this evidence with the disappearance of the ship, it was concluded that the wreck was indeed that of the *Clara Nevada*.

No one knew how many people perished in the disaster—the passenger list had gone down with the ship. There was no backup paperwork, either. In the wreck report, the president of the Pacific & Alaska Transportation Company stated that he had "no knowledge who was on board." This was hardly unusual. This was the beginning of a gold rush, during which emphasis was on making money and not on keeping records. As a result, passenger records were sparse. It could be assumed that anyone who wanted to head south and had passage money was taken aboard. Only one body associated with the *Clara Nevada* was ever found.[17] In March 1898, divers were sent to Eldred Rock to retrieve bodies. There is no record that any were ever recovered.

Estimates of the dead ran from a low of thirty to well over one hundred, in addition to the crew of thirty. While there is no way to come up with an accurate count, combining the list of the dead from all newspaper articles, there were at least forty-six individuals aboard identifiable by name and one corpse. But several newspapers stated a handful of women were onboard, for whom no names were given, as well as seven Klondikers, again with no names. There is also evidence to support the contention that there were at least three stowaways on board. Probably, the best count would be in the neighborhood of fifty-seven.

It did not take long for recriminations to begin. Every incident in the short life of the *Clara Nevada* was examined to see if a clue to the ship's demise could be discovered. Investigators reexamined the incident of January 27, when the *Clara Nevada* collided with the United States Revenue cutter *Grant* in Seattle. While at the time this was considered nothing more than an embarrassment to a captain on his maiden voyage, after the *Clara Nevada* went down, it was assumed to be positive proof of the captain's incompetence.

Investigation of this incident revealed little. The collision had been "gentle," and the impact had been so slight it "did not cause any cessation of the waving of adieus." Of possible greater significance, it was revealed at the hearings that the engine room telegraph cable had been broken prior to leaving Seattle and had not been repaired before the incident at the Port Townsend dock.

Passengers on the trip north did not have kind words for either the ship or crew. Some claimed that the crew was incompetent, "intoxicated" or both, and the ship unseaworthy. J.W. Lansing, M.D., reported that, among other indignities, passengers had to furnish their own candles with which to read and that those candles had been stolen by the crew "as soon as we turned our back." Further, food on board "was not fit for humans to eat. Before reaching Skagway our tea and coffee was made of salt water from the ocean. Part of the time we had no water to drink or to wash in."

Another passenger, Colonel Fred Wilson, "one of the most popular singers of mistrelsy [*sic*] in the Eastern States," was quoted as saying that the trip north had been so bad that "a meeting of the passengers was held on the deck and the prevailing sentiment was in favor of leaving the ship." Wilson convinced the passengers to stay on board for the remainder of the trip, "but it was a stormy one."

This story was confirmed by another passenger, Charles Jones of the Dalles, Oregon, who recalled that "a petition was prepared by the passengers and presented to the customs official [in Port Townsend] for the purpose of having the *Nevada* held so that [the passengers] could be transferred to another boat." Jones was also quoted as saying that

> *I was afraid the* Clara Nevada *would be wrecked from the time she left Seattle until Skagway was reached. We smashed into the revenue cutter* Grant *when we were backing out of Yesler's Dock; we rammed into almost every wharf at which we tried to land; we blew out three* [boiler] *flues; we floundered around in rough waters until all the passengers were scared almost to death; we witnessed intoxication among the officers and heard them cursing each other until it was sickening.*

Jones and R.C. Smelcer, another stampeder, both concurred that the voyage up had been one of "excitement and danger." The crew was not given high marks by the passengers either. Jones stated

> *Two-thirds of them were drunk. The second mate was put in penance twenty-four hours commencing Friday morning after we left Port Townsend. The first mate was full the night we left Seattle. He drank all the time but was yet able to be around and issue orders. The Steward was drunk all the time...I heard that the engineers were drunk. The freight clerk told me that the first engineer was taken off because he was a mason and not because he was a competent man. The freight clerk told me distinctly that as a matter of fact the engineer was incompetent.*

The longer the investigation of the maiden voyage of *Clara Nevada* continued the more damning evidence came to light. Before the vessel arrived in Port Simpson, British Columbia, it was reported that "she blew out three boiler tubes." Between Fort Wrangell and Juneau, "her bunkers caught fire." Hardly encouraging were the words of Captain Lewis, who was alleged to have declared that had the *Clara Nevada* been twenty more minutes in reaching Juneau, "all would have gone to hell."

Captain Lewis must have been quite concerned as to the safety of his ship, particularly with such a rambunctious crew on board. In addition to plowing into the Revenue cutter and striking the dock at Port Townsend, the boilers were causing trouble, as were the crew and passengers. The ship's internal communications system was down, and when the vessel arrived in Skagway, Lewis "yelled his orders to a man on the deck who yelled to the first engineer who yelled to the second engineer." For good reason, the maiden voyage of the *Clara Nevada* caused the captain to lose sleep. He was apparently so worried that he "did not dare go to bed. He was up day and night and always on the bridge."

The weather compounded human proclivity and error. Going north, the *Clara Nevada* fought the wind. Passage was slow, and at times, the ship made no headway at all. "On one occasion," one passenger recalled, the ship "did not make a mile in six hours."

Since there had been an explosion, the investigation naturally focused on the boilers. Suspicion of the boilers deepened when a former worker on the *Clara Nevada* claimed that while he was working on board he knew the "boilers were unsafe. Had I known the condition of them prior to sailing I would not have allowed my wife to sail on the ship, nor would I have done so, either."

Even the passengers knew there were difficulties with the boilers. This was corroborated by three passengers—T.L. Cockrill of Seattle, A. Boskey of Texas and Frank S. Duff of Marysville, Montana—who "said the engineer was having troubles with the boilers, and at Juneau the steamer was detained overnight while repairs were being made." One of the McGuire brothers, the owners of the *Clara Nevada*, was interviewed by the *Post-Intelligencer* and asked if this was an accurate statement. McGuire confirmed that it was.

McGuire further stated that he had received a letter from the engineer in Juneau, who stated that he had "found it necessary to put packing around certain tubes but this was a trifling matter and should occasion no uneasiness whatever regarding the steamer's condition." In the previous day's edition, the *Post-Intelligencer* had given a lengthy discussion of the state of the boilers. They had been examined by United States steamboat inspectors. One of the inspectors, Cherry, had even gone "into the boilers and inspected each thoroughly from top to bottom." McGuire also revealed that the *Clara Nevada* had been "carefully inspected by Captain Burnes and Captain Cliff, marine surveyors," before a $30,000 insurance policy was approved. According to H.P. McGuire, the *Clara Nevada* was insured at "1¼ percent, less than any other steamer running north, which is positive evidence of their entire satisfaction as to the *Clara Nevada*'s good condition."

Colonel Alden Blethen, publisher and editor of the *Seattle Times*, did not see it that way. On February 21, the *Seattle Times* reprinted a letter from "Observer," who was alleged to have been employed on the *Hassler* before she was transformed into the *Clara Nevada*. Observer stated that the "machinists who gave [the *Clara Nevada*] the overhaul freely expressed their opinion to the other workmen aboard that the *Hassler*'s boilers would never pass an official inspection, as she was liable to blow up at any moment if she was allowed to go into service." Competent firemen clearly knew this, according to an unnamed source quoted in the *Victoria Daily Colonist*, for they refused to serve on the ship, fearing being "scalded to death."

By the end of February, another matter came to the attention of the reading public. According to a number of sources, it was estimated that about $165,000—$13 million in 2011 dollars—in untraceable raw gold had gone down with the *Clara Nevada*. (For a real-world understanding of how much money this was in 1898, consider that a miner making $0.50 a pan was a worthwhile take.)

There were a number of published sources for the statement. The first was the February 25 *Seattle Times*, which estimated the gold at "$90,000 to $120,000." The June 5 *Post-Intelligencer* listed the gold as amounting to "$100,000." A *Seattle Times* article even indicated that one passenger was carrying $165,000 on him. The *Victoria Daily Citizen* (Victoria, British Columbia) did not list an amount but noted that "there is good reason to believe that a considerable sum in treasure was received on board at Skagway."[18] Other articles in the Seattle papers listed the amount of gold as high as $300,000. For the purposes of this book, the figure "$165,000 in gold dust and certified checks" was chosen as the amount as it comes from a reliable source and is between the highest or lowest estimates.

The thought of gold being on board is intriguing, but it was unusual as well. First, it was the wrong time of year for gold to be coming out of Skagway. Miners worked the Klondike while the rivers were running, and if they were heading south, they would have left when the rivers began to freeze in the fall. The *Clara Nevada* sank in February, long after freeze up. Further, most of the gold from Dawson was shipped out via the Yukon and St. Michael and not hiked to Skagway or Dyea and then sent out on a gypsy steamer.

Second, if indeed there was gold onboard, it was not institutional gold. In other words, it was gold owned by individuals—Klondike miners who had struck it rich—and not banks or corporations like Wells Fargo or the Canadian Bank of Commerce. These companies would not have risked sending their gold on what was, in essence, a tramp steamer. They would have depended on a more reliable ship with a schedule, like the *Queen*, *Islander*, *Tees* or the *Amur*. Further, the fact that no official showed up *after* the *Clara Nevada* went down is a good indication that the gold was owned by individuals, not a corporate entity.

Chapter 3

The *Seattle Times* and *Post–Intelligencer* Do Battle

The wreck of the *Clara Nevada* was unquestionably a disaster, but the tragedy was not without its silver lining. More than a few mariners in Seattle were appalled at what was happening to the reputation of their city generally and their industry specifically. Dollars were there for the reaping, and scoundrels were lining up to fleece the unsuspecting stampeders who packed the docks. Anyone with a few thousand dollars who found a derelict vessel could enter the shipping business. More than a few did. Unsavory crews of questionable ability were easy to find, and passengers were standing in line to pay for passage north. The maritime business had sunk so low that how long companies remained in business depended more on weather than the ability or competence of the captain and crew.

While there were Steamboat Inspection Service inspectors in Seattle, there were none covering Alaskan waters specifically. The Alaskan branch of the service began its work in May 1898, but by then, more than a dozen ships had already gone to the bottom of the Inside Passage. Conditions continued to be intolerable, and transportation north to the Klondike continued to be uncertain. By the end of the year, so many gold rush ships had gone down that the *Post-Intelligencer* was recording maritime disasters as fillers rather than front-page features.

Even with the Alaskan branch of the service in operation, the state of the shipping industry in Alaskan waters did not improve much. It did not affect the Seattle to Skagway traffic at all. The first commissioner concentrated on the maritime fleet into and out of "Yakutat, Nutchek, Virgin Bay, Valdez, Kodiak, Unga, Unalaska" and St. Michael. The Skagway area was not even visited until April 28, 1901, more than three years after the *Clara Nevada* had gone down.

In Seattle, the slowness with which the Steamboat Inspection Service was moving was not missed by the *Seattle Times*, and the continuing disasters in the shipping industry led to more than just recriminations. They precipitated an all-out journalistic war between the two powerful Seattle papers, the *Post-Intelligencer* and the *Seattle Times*.

The more opinionated of the two papers was the *Seattle Times*, published and edited by the acerbic colonel Alden J. Blethen. Described as both brilliant and controversial, Blethen had come to Seattle in 1896 and began his publishing career with the acquisition of a bankrupt paper, the *Press-Seattle Times*. He dropped the first half of the title, removed the old editor and, within six months, had more than doubled the paper's circulation.

But he had competition. When he had bought the *Press-Seattle Times*, there had been no other viable newspaper in Seattle. The *Post-Intelligencer* had been around since 1867 under a variety of names and publishers. In 1898, it had an institutional Republican bent, as opposed to Blethen's populist views, and the two came to journalistic blows frequently.

A rather large and juicy bone of contention between the two papers was the United States Steamboat Inspection Service. The *Post-Intelligencer* may have started the battle by commenting favorably on the service. Or maybe the *Post-Intelligencer* just needed to comment on the service to get a reaction from Blethen. Regardless of who lit the fuse, the powder keg blew. A significant milestone in the confrontation was published on January 22, 1898, in the *Post-Intelligencer*.

With all of the vessels now in Puget Sound, most of them on their way to Alaska, the *Post-Intelligencer* noted that

> *the weight of labor and responsibility borne by the board of inspectors has greatly increased. It is contended that it would be strange if some of those who have embarked in the Alaska transportation business did not overlook in their desire to reap a harvest from the Klondike trade the weakness or defects of some of their vessels.*

The article concluded with a quote by Captain W.J. Bryant, one of the two inspectors, stating that the policy was to be "strict and exacting but within the bounds of reason." Two weeks later, the *Clara Nevada* was at the bottom of Lynn Canal, and Colonel Blethen had a crusade.

It should not be surprising that the *Clara Nevada* became the center of a journalistic showdown between the two Seattle papers. It was grist for the editorial mill. The acid pen of Blethen quickly seized on the sinking of the *Clara Nevada* as an excuse to attack the Steamboat Inspection Service. The hidden motive for his continuing editorial diatribes was to embarrass the state of Washington's Republican administration, which was receiving favorable press in the *Post-Intelligencer*, his editorial archenemy. Though his attack may well have been journalistically motivated, the board of inspectors was a fair target, and Blethen's facts were accurate.

Moreover, many in the shipping business were probably more than pleased with Blethen's attack. They felt just as Blethen did, though they were by no means as vocal. Fearful that the good name of their city, a growing maritime power, was being tarnished by fly-by-night operators, Blethen's editorial opinions were welcomed. After all, those familiar with economics and reputations knew that the gold rush boom in shipping goods and passengers would pass and probably quickly. Once the brouhaha was over, they feared Seattle would be saddled with a reputation for placing profit before safety. That would leave a bitter taste in the mouth of serious shippers who would be doing business long after the gold rush was a dim memory.

Blethen was quick to the bulwark in charging that Seattle was acquiring a reputation as a "haven for decrepit and unsafe ships" and men who were more interested in profits than human lives. The flood of humanity to the gold fields seemed to bring out the wolf in some maritime companies, and more disasters were destined to follow the *Clara Nevada* incident, numerous Blethen columns claimed. The *New York Times* agreed with Blethen and noted that the *Clara Nevada* incident "simply emphasizes the conditions that prevail in Northern waters. Ships of all sorts and conditions are being pressed into the service to carry crowds to the gold fields."[19]

When the story of the *Clara Nevada* first broke, Blethen's vitriolic point of view was clear from the first paragraph of the first story. His editorializing was at its finest. Commenting on the trip north, he stated that 150 passengers and "an immense amount of valuable merchandise left this port in an unsafe vessel and in charge of a drunken and blasphemous crew...That she ever reached her destination is one of those modern miracles which God sometimes works in spite of men's failings, avarice, incompetence and greed."

Blethen further charged that a "blush of shame" should come upon the cheeks of the owners of the *Clara Nevada* and bring the "righteous indignation of an outraged public upon the heads of the culpable inspectors of the port." Here then was Blethen's call to arms against the inspectors and, consequently, the *Post-Intelligencer.*

If this was a glancing blow, Blethen finished with a barrage:

> *Once more the* [Seattle] Times *demands that the inspectors speak. You, Mr. Cherry, and you, Mr. Bryant. You who permitted that vessel to leave port—what have you to say? The public want*[s] *to know. The children, wives and mothers of those who were so suddenly ushered into eternity because you appear to have neglected your sworn duty want to know. Speak, men, speak or else by your moody silence you admit that there is some foundation for the odious suspicions now hurled toward you by the public you once swore to serve faithfully as well.*
>
> *Have you fulfilled that oath?*[20]

Throughout the spring, as hearings were held into the circumstances of the destruction of the *Clara Nevada*, Blethen charged the Steamboat Inspection Service with incompetence. The *Seattle Times* withstood a "vicious attack" by the *Post-Intelligencer* for its "expos[ure] of the manner in which the local inspectors have been conducting or, rather, neglecting the inspection of vessels at this port." Then, in the same article, it gave a laundry list of deficiencies, listed by ship, allowed to pass by the inspectors. Those ships included the *Clara Nevada*, *Eliza Anderson*, *Al-Ki*, *Mexico*, *Corona*, *North Pacific* and the *Alice Blanchard*.[21] On March 5, the *Seattle Times* went so far as to suggest the steamboat inspectors "decorate the end of an elevated rope." The head of the agency was accused of being "as fit for his office of Inspector General as a chimpanzee is to run a dime store."

If the *Seattle Times* had been looking for journalistic fodder, the investigation into the unsavory facts of the last trip of the *Clara Nevada* brought many to light. Incredulously, it was revealed, some life preservers placed on the *Clara Nevada* had been ones that had been on the *Hassler* and had been made of "grass." Though the McGuire brothers claimed that all grass life preservers had been removed and cork ones added, they could not account for the grass preservers that were found twisted in the wreck's debris.[22] In this, at least, the *Seattle Times* gloated. A government official had actually done his job! "The Custom House officials…refused to give [the *Clara Nevada*] clearance papers until she was provided with sufficient life preservers."

A week later, the *Seattle Times* was again attacking the "ex-fireman and his burly associate who permitted that floating coffin to leave this port." In a reprint from the *Victoria Colonist*, Fred Emery, supposedly one of the crew members who had gone down with the *Clara Nevada*, turned up alive. He was living in Dyea, where he was working with a Boston party to build a hotel.

According to Emery, the majority of the crew was "drunk and fighting the greater part of the time, the rowdy element being so conspicuous that the steward, Dan O'Connell, would have been thrown overboard on the up trip had not the captain interfered."[23]

Since witnesses reported seeing a fireball, it was generally assumed—and confirmed—that dynamite had been transported along with passengers, a clear violation of good sense not to mention maritime law. When the *Seattle Times* heard of the transportation of dynamite on a passenger ship, its journalistic acid was unrestrained. In a news article, Blethen proselytized that

> *the* Seattle Times *is not willing to believe that any creature made by God in the image of man could so harden his conscience as to load dynamite upon a vessel upon which many passengers were booked and send them all out of port with a full knowledge on his part that the slightest accident would land them all in eternity.*[24]

The dynamite on board the *Clara Nevada* had been destined for the Treadwell Mine in Douglas, across the Gastineau Channel from Juneau. Apparently it had been loaded first in Seattle and was thus at the bottom of the hold. To have offloaded the dynamite in Juneau would have required the crew to remove all the cargo for Skagway and Dyea before it could reach the shipment for Douglas. However, since no one was available in Juneau to receive the dynamite when the ship docked, it must have been decided to steam on to Skagway and Dyea, a day away, and then drop the dynamite off two days later when the ship had landed its Skagway cargo.

There is another theory, however, as to why the dynamite was not loaded. At least one witness felt that the dynamite had been "smuggled aboard and covered with the cargo." This was probably done because the captain and company owners knew full well that the transportation of dynamite with passengers was illegal. Then, when the vessel arrived in Juneau, the crew did not unload the dynamite because of the presence of a customs inspector on the docks. In any case, the dynamite was left on board when the ship headed north to Skagway and Dyea.

But there was more. That the *Clara Nevada* was poorly renovated, there was no dispute. It had been an old ship when it was sold for scrap, and the refitting had done little to ensure the seaworthiness of the vessel. But, as more and more details of the renovation came out, many mariners had cause to wonder how the vessel floated at all. Latty Boyce, the ship's carpenter who had fortuitously gotten off the ship in Skagway, wrote that "the ports and deadlights were not at all [caulked] before we left port and some never were [caulked] at all, while others had no threads in the bolts to screw them in so, had to be 'tomed' or blocked in with a brace of wood [*sic*]."

Once it was revealed that there had been dynamite on board, there was doubt as to what had caused the explosion. There was undeniably a "large hole, blacked and charred in the middle of the hull," which many would say was mute proof of the "awful tale of the boiler explosion."

Other mariners were not so certain of this conclusion. Why should the hole in the hull be "black" if there had been a boiler explosion? A blackened hole in the hull was more likely to occur with dynamite ignition. Further, if the witnesses had indeed seen a fireball, then the ship must have been ripped apart by an explosion of fire, not steam. A boiler explosion would not have caused a fireball that was immediately extinguished.

It was also postulated that it was possible that the wood around the boiler had caught on fire "not so much from any overheating of the boilers as from improper protection that should have attracted the inspector's notice."[25] This, the *Seattle Times* asserted, was probably the trouble discussed in the dispatch from the *Clara Nevada*. Two days later, almost as if in rebuttal, the *Post-Intelligencer* stated that "speculation as to the cause of the fire and of the explosion is unsatisfactory. It is asserted that there was not enough woodwork around the boilers to cause them to explode, even if the woodwork had caught fire."

Two days later, on February 17, the *Seattle Times* went after the inspection service again: "Competent and capable men only should be permitted to hold the position of steam vessel inspectors, and however competent the present incumbents of these offices may be to run a canal boat there is no question but that they have forfeited the confidence of a large number of people in this community."

Hearings on the disaster lasted until August 1898, when the conclusion was reached by the Steamboat Inspection Service that the ship had caught fire, "and during the frantic fight to keep the flames from the place where was stored powder and dynamite the officers lost their bearings and, incidentally, control of the ship." The storm drove the vessel onto the reef at Eldred Rock, "broadside on," where the ship split open. The primary evidence for this fire theory was that fire hose was found on the ship's deck, and the hose was "attached to the hydrants and coupled to the pumps."

Then the story of the *Clara Nevada* faded. But not without a last barrage from the *Seattle Times*. In disgust, Blethen insinuated that "criminals seem to have sufficient 'pull' with the powers that be to prevent an investigation until time and the elements have destroyed the last vestige of the evidence of their guilt." This charge was repeated verbatim on two occasions two days apart. The *Post-Intelligencer* did not rise to the bait.[26]

Though scrapping between the two papers continued for years, by April, the *Clara Nevada* had outlived its journalistic usefulness. Seattle was in the grip of a savage attack of Klondicitis, and the emphasis was on making money, not investigating mishaps.

There was another reason for the *Clara Nevada* to be forgotten. Ten days after the vessel had gone to the bottom of Lynn Canal, the *Maine* was sunk in Havana Harbor. War was declared by Spain on April 24 and the United States on the next day. American naval forces annihilated the Spanish force in Manila Bay on May 1. Marines landed at Guantanamo in Cuba on June 11 and repulsed a major Spanish assault on June 15, the same day Hawaii was annexed. The Spanish surrendered Guam on June 21, and the Spanish fleet at Cuba was destroyed on July 3. On July 25, Puerto Rico was invaded by the United States, and a Spanish assault on American troops near Manila was blunted on July 31. The breaking international news swamped the papers, and the saga of the *Clara Nevada* lost its power to generate headlines.

Additionally, as the flow of stampeders north grew from a trickle to a river to a flood, the number of maritime sinkings rose dramatically. In 1896, there had only been three marine wrecks. There were eight the next year. By the end of 1898, there had been thirty-four. With regard to the ships designed for Yukon River travel, the *Scientific American* reported, only five were still operating from the forty that had been built.[27] When recriminations ceased between the *Seattle Times* and *Post-Intelligencer*, the *Clara Nevada* became just another ship lost at sea.

But what became singular about the *Clara Nevada* was its revivification. Ten years later, almost to the day, another ferocious storm boiled the waters of the Lynn Canal. Ships scattered for the cover of bays and bights, and on Eldred Rock, the lighthouse keeper could feel the earth shifting beneath the tower. Though it was almost brand new, built in 1902, the structure shook violently. All night long the wind screamed as it powered its way down the channel. Waves rose to staggering heights and threatened to sweep across the small island.

It was not until morning that the wind died and the waves settled. Only then did the lighthouse keeper venture outside. What a sight he saw! There at the northern end of the island was the *Clara Nevada*, the ship and the cadaverous remains of her crew, sitting on the beach. The ocean had dragged the grisly ship out of its grave and up on the reef. Later that night, the same storm that had brought the ship up took her back to her watery grave.

This is a very fine story and was reported in similar language in Gordon Newell's book *SOS North Pacific*. Historical evidence, however, tells the story a bit differently. While the corpse of the *Clara Nevada* made one last appearance, it did not do so in such a dramatic manner. On February 25, 1908—just about ten years to the day later—M.J. Currie, keeper for the recently constructed lighthouse on Eldred Rock, reported that the *Clara Nevada* had "appeared in view off the north coast of the reef side of Elder [*sic*] rock, and that many bones of those who were long on the ill-fated vessel have been picked from the sea."[28] He felt that the "terrific winds" and "an extreme heap [*sic*] tide" had allowed the wreck to be seen. The sighting, however, was brief. The next morning, the wreck disappeared from sight.

When Gordon Newell was contacted and shown the discrepancy between his script and historical fact, he noted that "*SOS North Pacific* was written about 35 years ago, when I was a lot younger, more naive and given to accepting the tall tale of old salts at face value. The one provided in the case of the *Clara Nevada* was, to put it mildly, highly fictionalized."[29]

Today, what is left of the *Clara Nevada* sits in very shallow water to the north of Eldred Rock. Divers who have been on the craft state that at low tide, part of the ship is barely six feet beneath the surface. The superstructure is completely torn apart. James Nelson, a wreck diver out of Juneau, reported in 1991 that the *Clara Nevada* was "the most destroyed wreck [he] had ever seen," but it did have one highly visible landmark: the copper drive shaft for the propeller shaft that ran the length of the corpse of the *Clara Nevada*.[30] By 1993, that drive shaft had also disappeared, probably taken by salvage divers.

Chapter 4

"Curiouser and Curiouser"

Though the *Clara Nevada* settled comfortably into the status of a colorful incident in the history of the Klondike gold rush, there were too many inconsistencies for it to remain unexamined forever. It was a story that could not die and continued, as *Alice in Wonderland* said, "curiouser and curiouser" as the years went by.

Scrambling the chronology, in June 1916, eighteen years after the ship went down, Alaskan hard-hat diver C.F. Stagger spent two days on the wreck. In addition to cutting and "farming the kelp" that entombed the vessel in a living shroud, he also salvaged about half a ton of copper and brass. Though he could not have made it below decks, he was "positive from the examination made that the vessel had not caught on fire as was generally supposed and that the wreck was caused by something else, most likely a submerged rock."[31]

How accurate his statement was, after the ship had been under water for almost two decades and stripped for salvage at least once earlier, is unknown. But it does raise questions; if the ship had not burned, then why had there been a fireball?

The fireball theory is not without its flaws. It only fits if the witness or witnesses really did see such a phenomenon. For this to be true, a witness would have seen an incident that lasted, at the longest, two or three seconds at a distance of at least seven miles away in a raging storm. That seems a bit farfetched.

Another bit of evidence that also seems farfetched was printed in the *Post-Intelligencer.* According to the newspaper, even though there was a ninety-knot wind and a blasting snowstorm, "the long wharf at [Comet] was crowded with spectators" who heard a "loud report which resembled the explosion of boilers and nothing more was seen of the ill-fated steamer." If this story is accurate, the explosion had to have been extremely loud to be heard from seven miles away above the screeching of a ninety-knot wind.

Then, in the same paper on the same day, on the same page, is a story that describes the ferocity of the storm:

> *A terrible blizzard has been raging along the entire coast, from the head of Lynn canal to Fort Wrangell. So severe was the wind that vessels in a harbor were afraid to venture out. While on her way to Juneau the steamer* Chilkat, *from San Francisco, was caught in the storm. For ten hours she steamed against the wind, making but four miles…It is generally admitted that the storm was the worst in years. For a time, snow descended,* [and] *the cold was intense.*

A bit closer to the site of the wreck, the *Alaska Searchlight* in Juneau reported that "the wind was blowing a perfect hurricane from the north." These conditions were thus those in which "the long wharf at [Comet] was crowded with spectators." Further, the water off the Comet dock is very deep, so the dock was very short. The only dock in the area long enough to be "crowded with spectators" was at Jualin, in Berners Bay, on the inland side of a peninsula and, thus, out of sight of Eldred Rock.

As a tidbit destined for some book on the ironies of history, the only person identified by name who actually witnessed the fireball was "George Beck." It is assumed that there is no relationship between this man and the George Foster Beck who died that night on the deck of the *Clara Nevada* and whose body would be the only one ever recovered.

But there was more evidence supporting the fireball theory. In March 1898, the *Dyea Trail* reported that an investigator at the site, Sanderson Reed, believed that the *Clara Nevada* had gone aground, and that fire might have occurred when lamps were tipped over as the *Clara Nevada* struck Eldred Rock. "The fire evidently started," he surmised, "in different parts of the vessel, which prevented the crew and passengers from checking it." This, however, would not account for the large explosion heard as far away as Comet.

Latty Boyce, the *Clara Nevada*'s carpenter who got off the ship in Alaska confirmed this speculation: "The saloon [of the *Clara Nevada*] was lighted by four large lamps set loosely on the tables and I saw one roll over and break the shade myself. I thought them very dangerous." However, since the letter from Boyce appeared a week after the original statement from Sanderson Reed, it could be assumed that Boyce was responding to Reed's speculation.

Reed also noticed a large hole in the side of the vessel but believed the boilers to be intact. He guessed that the striking of the reef had caused fires to erupt in different parts of the ship, making it impossible for the passengers to fight all of them at the same time. This corroborated testimony by a Lloyds of London surveyor, who stated categorically that there had been no boiler explosion.

Interestingly, Reed noted that there had probably been an attempt to lower the lifeboats, "which was unsuccessful on account of the storm."[32] This last statement is enigmatic, for it opens to the door to speculation as to how many lifeboats were indeed lowered, an issue that would haunt the issue of the last moments of the *Clara Nevada*.

While there is no question that the *Clara Nevada* did sink, events postdating the wreck add a strain of the macabre to the story. The *Daily Alaska Dispatch* reported that a human skull had been found in Auk Bay, just outside of Juneau. The skull, the article claimed, was undoubtedly that of one of the unfortunate passengers of the *Clara Nevada*, which had "struck a reef outside of Auk Bay in 1898 and afterwards burned to the water's edge with all hands lost." This editor was probably new to the job and may not have known that it has been firmly established where the *Clara Nevada* had gone down but not why.

Of greater interest, however, was a tidbit on the body found by the *Rustler*, allegedly that of George Foster Beck. According to the *Dispatch*,

> *the purser of the steamer* [George Foster Beck] *was the son of a wealthy family of Portland, Oregon. Thousands of dollars offered for the body eventually recovered what was purported to be the body. However, it has been often claimed to relieve the anxiety of the aged mother a fake body was shipped to the family burying ground at Portland and interred.*[33]

While one news article said there were rumors of other bodies being found, other than the skull that allegedly came from a victim of the wreck, no other recovery of bodies was mentioned in the newspapers. This raises an obvious question: if there were as many as 140 people on board—depending on whose numbers are used—how is it possible that a steamship could find only 1 body, and that 1 body had a reward on its recovery?

Twenty years later, the *Princess Sophia* went aground on Vanderbilt Reef fifteen miles to the south of Eldred Rock. With a heavy wind and a rising tide, the ship's keel was scraped open, and the ship sank in water so shallow its spars were above the tide line. More than three hundred passengers and crew went down with the ship, and their bodies littered the shores of Lynn Canal for miles. The owners of the *Princess Sophia* also offered rewards for the return of bodies, which generated a great deal of small boat activity in the area. With the *Clara Nevada*, the owners made no such offer, and thus there was not a frenzied search for bodies. Only one body was recovered from the *Clara Nevada*. It has to be assumed that the rest were entombed in the ship when it went down.

Further muddying the waters of investigation is an article that appeared in the *Dyea Trail* and *Alaska Miner* (Juneau) in April 1898. The identical articles reported that the steamer *Seaolin*[34] had spotted a boat on the beach near Comet, several miles south of Eldred Rock. It sent a crew ashore, and they found a boat that was "neither a lifeboat nor a sealing boat." There was a life preserver from the *Clara Nevada* on board, as well as a roll of blankets containing clothing, "which a Yukoner would take along with him."

Another blanket roll with clothing was found about sixty yards away and close by were the remains of a fire. The life preserver was singed, leading the rescue party to conclude that the *Clara Nevada* had indeed burned.

Opening the blanket rolls, the crew found a light check suit with the name "Geo. Kasey" in the pocket. Other papers found related to the purchase of goods by "William Hemming, Rockport, Ind." and a memorandum signed "Logsdon." The accounts further stated that "no trace of the bodies could be found. Everything point[ed] to the loss of two lives." No reason for this last assumption was given.

Another oddity of the Alaskan papers was the apparent lack of mention of the gold onboard. Considering that a fortune of gold could very well have been sitting in twenty-four feet of water, there were only a few recorded cases in 1898 of anyone attempting to salvage the ship. The first was on June 5. A tug named *Scow* was reportedly busy at work, trying to retrieve whatever gold was left after the wreck. A week later, there was another diver searching the wreckage. G.W. Stowel made "a great many dips spending several hours under water each time." He walked from one end of the sunken wreck to the other several times and brought up quite a few artifacts that were subsequently put on exhibition "in Juneau stores." Later that year, in mid-August, there was a second report of activity. The firm of Beddoe & Nowell was at work, looking for the *Clara Nevada*'s safe. "From the position of the wreck," it was reported, "it is plain that the *Nevada* was split open and it is believed that the safe dropped through the ship into deeper water." While it was claimed in this article that there had been $300,000 in gold on board, there was never a report of any gold actually being raised.

Could there have been others who did their work secretly? Possibly. But it is important to note that the *Clara Nevada* sank in the middle of a narrow channel that was being used frequently by ships loaded with stampeders, making the run between Seattle and Skagway. It would have been hard to do the work on the sly. However, if anyone saw any salvage operations, it did not make any of the papers along the Lynn Canal or those in Seattle.

What happened to the gold?

Aerial view of Eldred Rock, unknown date. *Courtesy of the United States Coast Guard.*

Clara Nevada, as the United States Coast and Geodetic Survey vessel *Hassler. Courtesy of the Alaska State Library Robert DeArmond Collection 134-288:2.*

William H. Evans. Courtesy of the Alaska State Library DeArmond Collection 234-310-2.

The dock at Comet, Alaska, date unknown. *Courtesy of the United States Coast Guard.*

Traveling shows were the entertainment of the day. Note the variety of offerings at this Skagway theater. Also note that women were not excluded from such entertainment. *Courtesy of the Alaska State Library P226-842.*

Above: *Clara Nevada* with two support boats, probably in Skagway. *Courtesy of the Alaska State Library PCA-87.*

Left: Eldred Rock lighthouse and crew, unknown date. *Courtesy of the United States Coast Guard.*

Unknown roadhouse in Alaska. Note that all the men have slicked down their hair for the photograph. Also note the heavy boots on the men in the front. These were needed in the spring when the trails turned to mud. *Courtesy of the Anchorage Museum of History and Art B 81-19-154.*

The shoals of the Lynn Canal were equally dangerous for both cargo and passenger ships. Here, in 1910, the SS *Princess May* was left high and dry on Sentinel Island when the tide went out. *Courtesy of the Alaska State Library Norton Collection 226-614.*

The lighthouse on Eldred Rock was built in 1903 and is located between Juneau and Haines on the Lynn Canal. It is appropriately located because Eldred Rock, where the *Clara Nevada* went down, is the only spot in the Lynn Canal where the rocks break the surface of the water midchannel. It is, in essence, the top of a mountain that rises 1,500 feet from the depths of Lynn Canal. *Courtesy of author's personal collection.*

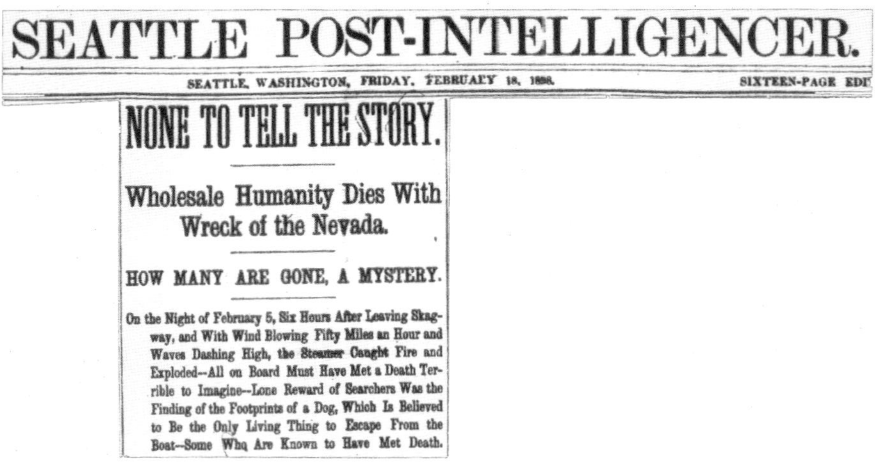

SEATTLE POST-INTELLIGENCER.

SEATTLE, WASHINGTON, FRIDAY, FEBRUARY 18, 1898. SIXTEEN-PAGE EDI

NONE TO TELL THE STORY.

Wholesale Humanity Dies With Wreck of the Nevada.

HOW MANY ARE GONE, A MYSTERY.

On the Night of February 5, Six Hours After Leaving Skagway, and With Wind Blowing Fifty Miles an Hour and Waves Dashing High, the Steamer Caught Fire and Exploded—All on Board Must Have Met a Death Terrible to Imagine—Lone Reward of Searchers Was the Finding of the Footprints of a Dog, Which Is Believed to Be the Only Living Thing to Escape From the Boat—Some Who Are Known to Have Met Death.

Headlines of the sinking of the *Clara Nevada*. *Courtesy of author's personal collection.*

This is the Candle Creek Roadhouse, date unknown. In the early days of Alaska, roadhouses were established a day's walk from each other. The food was good enough to keep you alive, and the liquor was usually illegal. *Courtesy of the University of Washington Nowell Collection 2246.*

Dog "sleds" were not just used during the winter. Here a dog "sled" is used as a water wagon in Nome, 1900. *Courtesy of the Anchorage Museum of History and Art B87.7.60.*

RNON, Ballard.

ings Bank has a
perty in various
oved and unim-
the orders of the
can be bought at
n easy terms. in
r will be refused.
y and full partic-
plication to
NG, Receiver,
Seattle, Wash.

RGAINS apply to
New York Block:
six rooms each, on
best locality of
2 per cent on in-
Eight-room house
led street, sewer,
investment; $1100
ence lot, 60x120 in
sirable; only $900.
your property.

these two lots in

ot 60x120, close in,

from car line, 3
nly.
, $125.
Columbia Street.

nd fenced, 7-room
0.
000.
om water, 30 acres
house, large or-
chicken house and

to 8 p. m. Decker Hotel, Seneca Street and First Avenue. Rooms 3, 4, 5 and 6.

ISMAR THE GYPSY, nature's greatest living clairvoyant and palmist, has returned to the city and can be consulted on all affairs of life; hours from 9 a. m. to 8 p. m. Parlors 1 and 2, Shorey Block, corner Third and Columbia. Fee $1 and up.

MME. LARABEE, the wonder of the age; greatest astrologist and palmist in the world; gives advice on all business; reads your life from the cradle to the grave; this lady possesses great healing power; also assists lost friends. 1506 First Avenue.

MRS. LENONT, reliable spiritual and business medium; information of gold localities by "White Bear," her Alaskan guide. Public circles under spirit control Monday, Thursday, Friday, 8 p. m. Second floor Safe Deposit Building, First Avenue.

MRS. E. EMMETT, the South African world-renowned medium, in her wonderful clairvoyant sight, reveals the past, present and future; consult her; readings, 50c. Rooms 22 and 23, 811 Second Avenue.

CONSULT RAYMOND, BOY MEDIUM, about Klondike; truthful predictions, independent slate writing; circles Wednesday, Friday and Sunday evenings. 1422 Second Avenue.

MRS. ESTHER THOMAS, Psychic—Harvard, corner Second and Virginia. Room

PACIFIC COAST S
FOR SAN FR
The Company's elega
TILLA, CITY OF PU
WALLA leave SEATT
send and Victoria, B.
Feb. 3, 8, 13, 18, 23, 28,
30, Apl. 4, 9, 14, 19, 2
FRANCISCO for Seat
C., and Port Townsen
Feb. 5, 10, 15, 20, 25, Ma
Apl. 1, 6, 11, 16, 21, 26.
FOR AL
The Steamers QUE
CITY OF TOPEKA,
CITY and CURACAO
a. m. Feb. 4, 7, 9, 14, 1
1, 9, 11, 12, 16, 19, 24,
10, 15, 17, 18, 20, 25, 28,
pany reserves right
previous notice steame
hours of sailing. For
procure company's fol
J. F. TROWBRIDGE
Dock, Seattle.
City Ticket Agt, 606
Goodall, Perkins & C

SEATTLE-TACC

Four Round Trips Da

Everyone tried to "make a buck" on both the Klondike and Alaska gold rushes. Here are some wily entrepreneurs advertising in the Seattle paper. *Courtesy of the Museum of History and Industry in Seattle LB 124087.*

Seaolin was supposed to have been named "*Sea Lion*," but the painter had allegedly visited a few saloons on his way to work. *Original source unknown, this photo appeared as a reprint in the* Alaska Sportsman, *May 1966.*

An authentic four-man Alaskan gold mining operation. Note the large steamer and the pipe that directs the steam into the ground. Because the ground was frozen, it had to be thawed before the gold could be extracted. The earth that was thawed was put in a spoil pile, which you can see on the left-hand side of the photograph. Then the spoil pile was washed one pan at a time in water from the creek, which can be seen on the far left side of the photograph. *Courtesy of the University of Washington Nowell Collection 143.*

Main Street in Nome in 1900. Note the Dexter on the left hand side of the photograph. That was Wyatt Earp's saloon. *Courtesy of Jim Green.*

Bessie's Roadhouse in Nome, presumably with Bessie in the wagon on the far left. Women were more than welcomed during the Alaska gold rush. That's because there were so few of them, about 10 percent of the population. Note the "Ladies Entrance" at the back right of the photograph. *Courtesy of the University of Washington Nowell Collection 6650.*

All kinds of cargo were transported to Dawson. Here, during the summer of 1898, a mule train takes crates of turkeys over the White Pass out of Skagway. *Courtesy of the University of Washington 8291.*

Cold weather did not stall business during the Alaska gold rush. *Courtesy of author's personal collection.*

Express dog teams from Iditarod in Seward. These sleds were for cargo, not mail. Note the length of the dogsled at the far left. *Courtesy of the Alaska Museum of History and Art B81.19.2.*

Here are the United States Army troops who kept order on the Alaska side of the border during the heady days of Klondike strike. This picture was taken in Skagway on July 4, 1899. Please note that these are all buffalo soldiers, black men, and it appears that the officers are black as well. *Courtesy of the Alaska State Library PCA-75-144.*

Clara Nevada on her only voyage. *Courtesy of the Jefferson County Historical Society, Port Townsend, Washington.*

The Washington-Alaska Bank in Fairbanks. It was the creation of E.T. Barnette, the man in the white hat in the center of the photograph. In 1911, Barnette absconded with all of the money in the bank. He eventually went to trial but was found not guilty. He lived the life of a lord in Mexico for a decade and then moved to Los Angeles, where he died under mysterious circumstances in 1933. *Courtesy of the University of Alaska Fairbanks 63-46-147N.*

Lighthouse duty has never been easy. Duty on them could last years, and the men who maintained these structures could go for months without seeing another soul. Eldred Rock is on a very small island, so small you can walk its perimeter is less than five minutes. Sometimes the waves of Lynn Canal are so high that you cannot walk on the island at all. This lighthouse was established in 1903 and was fully operational until the modern era. Today it is a visual reminder of where Alaska's ghost ship, the *Clara Nevada*, went down along with 165 people and 100,000 ounces of gold. *Courtesy of the United States Coast Guard.*

Above: Generally speaking, there were two types of paddle-wheelers, those with a single paddle wheel at the back, called stern-wheelers, and side-wheelers such as this one. The *Eliza Anderson* is famous for "the stranger who came aboard." Just as the ship was about to sink off the coast of Kodiak, a giant of a man broke into the wheelhouse and saved the ship. After the ship was safe in a protected cove, the man disappeared from the ship. Half a century later, a historian discovered documents that indicated the stranger was actually a stowaway, not a ghost. *Courtesy of the Alaska State Library DeArmond Collection 134-252-4.*

Left: Alden Blethen, editor of the *Seattle Times*. *Courtesy of the* Seattle Post-Intelligencer.

An authentic Alaska gold rush saloon. This one was in Koyukuk about 1900. Note the pinups of the day on the wall behind the bar. *Courtesy of the Alaska Museum of History and Art Wyman Collection 201.*

Photograph of the Alaska Commercial Company store in Arctic Village. Note the mastodon tusks, common discoveries in Alaska where men were digging up thousands of cubic feet of earth every year. *Courtesy of the Alaska State Library C.L. Andrew Collection 45-277.*

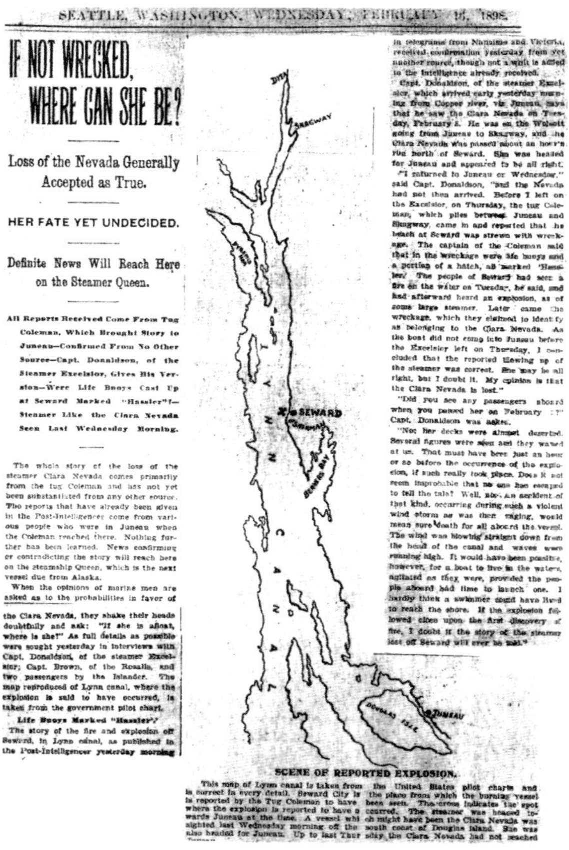

SEATTLE, WASHINGTON, WEDNESDAY, FEBRUARY 16, 1898.

IF NOT WRECKED, WHERE CAN SHE BE?

Loss of the Nevada Generally Accepted as True.

HER FATE YET UNDECIDED.

Definite News Will Reach Here on the Steamer Queen.

All Reports Received Come From Tug Coleman, Which Brought Story to Juneau—Confirmed From No Other Source—Capt. Donaldson, of the Steamer Excelsior, Gives His Version—Were Life Buoys Cast Up at Seward Marked "Hassler"?—Steamer Like the Clara Nevada Seen Last Wednesday Morning.

The whole story of the loss of the steamer Clara Nevada comes primarily from the tug Coleman and has not yet been substantiated from any other source. The reports that have already been given in the Post-Intelligencer come from various people who were in Juneau when the Coleman reached there. Nothing further has been learned. News confirming or contradicting the story will reach here on the steamship Queen, which is the next vessel due from Alaska.

When the opinions of marine men are asked as to the probabilities in favor of the Clara Nevada, they shake their heads doubtfully and ask: "If she is afloat, where is she?" As full details as possible were sought yesterday in interviews with Capt. Donaldson, of the steamer Excelsior; Capt. Brown, of the Rosalie, and two passengers by the Islander. The map reproduced of Lynn canal, where the explosion is said to have occurred, is taken from the government pilot chart.

Life Buoys Marked "Hassler."

The story of the fire and explosion off Seward, in Lynn canal, as published in the Post-Intelligencer yesterday morning in telegrams from Nanaimo and Victoria, received confirmation yesterday from yet another source, though not a whit is added to the intelligence already received.

Capt. Donaldson, of the steamer Excelsior, which arrived early yesterday morning from Copper river, via Juneau, says that he saw the Clara Nevada on Tuesday, February 8. He was on the Wolcott going from Juneau to Skagway, and the Clara Nevada was passed about an hour's run north of Seward. She was headed for Juneau and appeared to be all right.

"I returned to Juneau on Wednesday," said Capt. Donaldson, "and the Nevada had not then arrived. Before I left on the Excelsior, on Thursday, the tug Coleman, which plies between Juneau and Skagway, came in and reported that the beach at Seward was strewn with wreckage. The captain of the Coleman said that in the wreckage were life buoys and a portion of a hatch, all marked 'Hassler.' The people of Seward had seen a fire on the water on Tuesday, he said, and had afterward heard an explosion, as of some large steamer. Later came the wreckage, which they claimed to identify as belonging to the Clara Nevada. As the boat did not come into Juneau before the Excelsior left on Thursday, I concluded that the reported blowing up of the steamer was correct. She may be all right, but I doubt it. My opinion is that the Clara Nevada is lost."

"Did you see any passengers aboard when you passed her on February [illegible]?" Capt. Donaldson was asked.

"No; her decks were almost deserted. Several figures were seen and they waved at us. That must have been just an hour or so before the occurrence of the explosion, if such really took place. Does it not seem improbable that no one has escaped to tell the tale? Well, no. An accident of that kind, occurring during such a violent wind storm as was then raging, would mean sure death for all aboard the vessel. The wind was blowing straight down from the head of the canal and waves were running high. It would have been possible, however, for a boat to live in the waters, agitated as they were, provided the people aboard had time to launch one. I hardly think a swimmer could have lived to reach the shore. If the explosion followed close upon the first discovery of fire, I doubt if the story of the steamer lost off Seward will ever be told."

SCENE OF REPORTED EXPLOSION.

This map of Lynn canal is taken from the United States pilot charts and is correct in every detail. Seward City is the place from which the burning vessel is reported by the Tug Coleman to have been seen. The cross indicates the spot where the explosion is reported to have occurred. The steamer was headed towards Juneau at the time. A vessel which might have been the Clara Nevada was sighted last Wednesday morning off the south coast of Douglas island. She was also headed for Juneau. Up to last Thursday the Clara Nevada had not reached

Above: An authentic Alaska gold rush dance hall, not a roadhouse. This dance hall was in Arctic Circle, so-named because the boomtown founders thought the town was on the Arctic Circle. (It wasn't.) There are two women in the front, and if you look carefully, you will see a light bulb on the moose above the front door. Also note the boots the men are wearing—this was deep mud country. *Courtesy of the Anchorage Museum of History and Art Wyman Collection #202.*

Left: Map of the Lynn Canal and headline of the *Clara Nevada* missing. Seattle Times, *February 16, 1898.*

Teamsters in the days where "teamsters" meant men who drove horses. This photograph was taken in Nome in 1907. *Courtesy of the University of Washington Nowell Collection 5885.*

Not all dogsleds were pulled by dogs. Here is a "sheep" dogsled. Contemporary accounts stated that the sheep could pull as much weight as the dogs and were superior on the trail because they could forage for their own food. This meant the musher did not have to carry feed for the animals. *Courtesy of the University of Alaska Fairbanks Arne Wahte Collection 64-34-19.*

The loading of the steamships was functional but not elegant. *Courtesy of the Anchorage Museum of History and Art Messer Collection.*

This is a shot of incoming prospectors during the Sushanna Stampede in Cordova in 1913. From this photograph, it is clear that landing facilities had improved since the Klondike strike. Of interest because you can compare this photo to others, it is easy to see that these are stampeders coming in. Note their clean boots. *Courtesy of the Anchorage Museum of History and Art 377.5.22.*

An authentic Klondike gold rush cache. Men would transport their one thousand pounds of supplies one hundred pounds at a time. They would carry that one hundred pounds for two or three miles and put it in a pile. Then they would return for another load. After the one thousand pounds had been moved the two or three miles, the process would start over again. The word cache, which means to store something until you returned for it, is one of the few Alaskan words to make it into modern usage. Another is hooch, meaning a low quality of liquor that originally came from the Native American village of Hoochenoo located just outside of Juneau. *Courtesy of the Anchorage Museum of History and Art Messer Collection.*

This is a look at the legendary ice staircase of the Chilkoot Pass outside of Dyea. At the bottom of the staircase was Alaska; at the top, Canada. This is a significant historical photograph. Most historical texts on the Klondike gold rush state that the stampeders were constantly on the move up the staircase. It was either up or out. If you could not keep up with the pace of men lock-stepping up the staircase, you stepped off the staircase and slid to the bottom. This is supposedly the origin of the term "getting cold feet." However, note that there are at least three areas where men are resting alongside the staircase. Also note the half canoe on the back of the man in the foreground. Everything that had to go up the staircase was reduced to its smallest component, in this case half of a canoe. *Courtesy of the author's personal collection.*

Chapter 5

The Search for Answers

"History," I often tell students in my history class, "is not a re-creation of the past. It's an assessment of the past based on documents provided by people in archives and museums who will answer your letters." In piecing together the saga of the *Clara Nevada*, I learned just how true my statement was.

It did not take me long to realize that something was amiss with the accepted version of the wreck. Just a cursory examination of the facts suggested that something far more sinister than just a shipping disaster. There were too many loose ends.

Here was a ship with $165,000 in 1898 dollars, reputedly in raw gold, which sinks in twenty-four feet of water on one of the few reefs between Skagway and Juneau where the bottom usually plunges to depths of at least six hundred feet. That the gold was never recovered seemed odd. What happened to the gold?

In my quest to answer this question, I began to search the historical record for clues to a disaster that was almost one hundred years old. It meant going through all newspapers, archives, memoirs, diaries and letters of the era to discover what documents still remained. What I thought would be a relatively quick examination ended up taking more than three years, and even then I was unsure I had seen every relevant document still in existence. Such is the anxiety of the historian, the fearful belief that as soon as you complete your research and write your findings, someone will find a single document that refutes everything that you have concluded to be true.

I quickly discovered that the most reliable and unreliable sources at my disposal were the same: the press. Reading the papers still in existence, I found plenty of information, but much of it was contradictory. But that was the smallest of my headaches. It was nothing in comparison to the problems I had in tracing individuals. Names were spelled differently in different sources. Minor facts, such as the geographic location of evidence, were not in sync.

Take, for instance, the case of the discovery of the only body, that of George Foster Beck. I found three newspaper articles that stated the body had been found in three different areas, which stretched along a forty-mile front. Complicating matters, each of the three articles included critical information the other two did not. This meant that no paper was completely right, but then again, no paper was totally in error either.

(I also found an article in the *Morning Oregonian* a month after the sinking that two more bodies had been found, and that one was believed to be that of Captain Lewis. The two corpses were buried without their clothing being checked. No other source verified this story.)

Then there was the difficulty in actually reading the papers. Some of the newspapers I needed were in Alaska, but they did not constitute the bulk of my research. That came from the *Post-Intelligencer* and *Seattle Times*. As both of these were Seattle papers, I had to get the microfilm through interlibrary loan. When I started my examination, to my horror, I discovered that many of the pages were illegible, faint or smeared. This was as much a sin of printing as of microfilming. Some issues of the *Seattle Times* had dates of the previous day, and occasionally a paper was printed with no date. Tracking from secondary sources and the handful of indexes also proved to be extremely difficult, and often I found the cited March 23 item on March 10 or 15. Sometimes I did not find it at all.

Worse yet, 1898 was a long time ago, and many local newspapers for that era no longer exist. Often I found that critical days were missing; even more agonizing, sometimes it was months. I was particularly distressed to find that the microfilm of Juneau's *Alaska Miner*, a critical paper for the study of the *Clara Nevada*, had a heart-stopping, gut-wrenching gap from January 22 to March 12 and then *another* gap until April 9. The March 12 edition, incidentally, contained an article on the discovery of the body of George Foster Beck as well as an editorial on the *Clara Nevada*. Another Juneau paper, the *Searchlight*, had only a single issue still in existence in the critical time frame for the *Clara Nevada*: February 12. The third local paper I could have used was the *Daily Alaska Dispatch*, whose surviving issues started well after the *Clara Nevada* went down. If those specific papers exist, I have yet to find them.

With the Seattle papers, I was luckier. But then again, there were no indexes. Critical articles had to be discovered by reading the papers. This meant well over two hundred hours in front of a microfilm screen, scanning and hopefully finding every article and filler that existed.

This work is the culmination of that search through the dusty niches of historical documentation in the Pacific Northwest. It is a story based as much on luck as on knowing what I was doing. Unlike many historical works, this book is not based on a single body of documents with peripheral material added for spice. In fact, this entire work is one of peripheral material drawn from a wide variety of sources. I refer to this as scholarship by accident because how successful the writer is depends entirely on how clever he or she is at finding those documents that still exist.

My initial task was to search the historical record for any clues. I started with the United States census of 1900 for Alaska, Oregon and Washington, under the assumption that any survivor would most likely be a crew member and would still be in one of those three areas. A passenger who survived would have gone on to his final destination while a crew member would simply have signed on to another ship and would probably have been listed in some census in the Pacific Northwest. A search was made of the newspapers and every name identified with the *Clara Nevada* was listed.

However, using the 1900 census proved to be more difficult than spending an hour or so with an index of names. The census returns from Oregon and Washington, the most likely home ports for many of the mariners, were not indexed. Finding an individual meant I had to go through the returns of each county in each state and comparing each name to the list of forty-five that I had. This tedious task was made even worse by the fact that for some individuals I had no first names, and for others, different newspapers had spelled the last names differently. Further complicating the search, some of the individuals listed had first names that were probably nicknames, like "Bat" and "Peanuts."

Then there was the problem of reading the handwriting of the census takers. This was made infinitely more difficult by the fact that I was reading microfilm rather than the original documents. All of the data had been entered by hand, and each section of the census offered me the chance to interpret a new bit of penmanship. Page by page, the handwriting could vary. A poorly written "o" could be read as an "a" or an "e." Sometimes the writing was so light, it could not be read at all. Other times it was so dark that the ink had come through the paper and obscured the name on the next page.

Eventually I abandoned the search of the Oregon and Washington census data and concentrated on Alaska. Here I had an advantage: an index. But it was not a conventional index. Census data in those days was sequenced on the basis of sound and recorded in a "Soundex." Names were coded on the basis of how they were pronounced, not how they were spelled. This coding system, unfortunately, was difficult to interpret. As an example, take the name Hildebrand. A researcher would go to the "H" section of the Soundex because the word started with an "H." Since vowels had no assigned number, the "i" would be ignored. The next letter, "l," had an assigned number "4." The next consonant, "d," had the assigned number "3." The next letter, "e," would be skipped because it was a vowel and the final letter in the coding, "b," had a designation of "1." Thus, for the purposes of the 1900 census, "Hildebrand" would be found in an index as "H431."

Once the names of all passengers and crew had been assigned a Soundex number a search of the Alaskan census data was conducted. The problem now was to identify the exact name with one that had been listed in the newspaper. A close match would not do; it had to be an exact match.

When I finished, I had six possible matches for the crew in Alaska. Some, however, were quite speculative. Was Bat Hurley, the quartermaster on the *Clara Nevada*, for instance, "William Hurley," a waiter on the *City of Topeka*? There were, after all, two men on the *Clara Nevada* that were listed as quartermaster. One of those listings could have been a mistake.

Four "George Rogers" showed up on the census, one of them a seaman on the steamer *Alpha of Victoria B.C.* and the second as a waiter on the steamer *Tess*. The two others were a miner and a passenger. Making a match of George Rogers might be a moot point; however, some sources state that Rogers may not have been onboard when the *Clara Nevada* left Skagway.

Then there was Frank O'Donnell, the steward on the *Clara Nevada*. In this case, there was a close match. There was only one whose occupation was listed as seaman. He was on the steam whaler *Jeanette of San Francisco*. But there is no way to definitively state that "O'Donnell" with two "ls" is rare enough to assert that these two men were the same individual.

While "O'Donnell" might be unusual, other common names were not. In the case of Ed Kelly, pilot, and (no known first name) Kelly, waiter, of the *Clara Nevada*, there was no trouble finding similarly named individuals. Four references to "Charles Kelly," two for "Edward Kelly" and one for an unnamed Kelly exist in the record. Three of them lived in Skagway, the last port of call of the doomed steamer. In my opinion, none of these could be called a close match.

There was a sort-of match for Dan Cunningham. I found a "D. Cunningham" who was a passenger on a steam whaler *Thrasher* of New York. William Jacobs, carpenter, also has a name match with a boarder in Ketchikan, but these matches were doubtful. Jacobs, like Kelly, is too common a name to assume that these two men were the same individual.

With Paddy MacDonald, the fireman, the case for a match was closer, assuming that the census taker considered "MacDonald" and "McDonald" as the same surname. If "Paddy" was a nickname, there are three possible matches and one passenger who was obviously not the man in question. Two of the three contenders were waiters on steamers.

The third, William McDonald, was a forty-six-year-old fireman for the steamer *Humboldt*. He was living at "Moore's Wharf" in Skagway in 1900. It would be easy to jump to conclusions as this man's last name and profession are similar, and he was residing close to the location of the shipwreck. However, I discounted William MacDonald for two reasons. First, his name was not "Paddy," and that was the specific first name for which I was searching. Second, I found other references to the specific name "Paddy MacDonald."

The name Paddy MacDonald—recorded as McDonald in the Soundex—pops up a few more times in other historical sources. A man by that name appears in *Teller News* on August 8, 1901, as a "well known old timer." Then there is a Paddy MacDonald who was listed by the Royal Canadian Mounted Police as "one of the worst criminals ever known on this continent." On March 18, 1900, the *Nome Nugget* listed MacDonald and twenty other rogues who were assumed to be on their way to Nome from the Klondike gold fields. These men were considered a danger to any community they encountered and were, thus, men of which law enforcement officials should be wary. I also found an obituary for a Paddy "McDonald" in the May 31, 1935 *Anchorage Times*. Though this name is slightly misspelled, it is still very close to a match.[35]

Interestingly, with regard to Soapy Smith, the desperado who ran Skagway until July 8, 1898, one of the men taken prisoner after Soapy had been shot was W.F. Saportas. One of the passengers on the *Clara Nevada* was an E.W. Saportas. Those initials are close enough to suggest a match. However, a letter to the editor later revealed that these men were brothers.

Of no historical significance but a tidbit to titillate the imagination of the detective novel aficionado, there was a corpse on board. The "passenger," so to speak, was William Malloy. A search of the state of Alaska archives revealed no death certificate for such an individual, but this is hardly significant. There was no requirement for the issuance of such a certificate until 1913. In fact, there are no death certificates for any of the crew or passengers of the *Clara Nevada*. Malloy is interesting in that there are several references to a man of a similar name. One was a William J. Malloy, who was listed in the 1906 Fairbanks's Directory as being in "mining." Two of the men had what could be called death certificates. One of these was a "William Maloy, of La Conner," who was twenty-one, "a robust, healthy young man" who had died of pneumonia inland. He had only been in the northland for two weeks. His body had been brought down from Skagway by the *Rosalie*, presumably for transportation for burial in the United States. The other reference was to William Malloy, age twenty-two, who died of "cerebro-spinal meningitis" in the Skagway Hospital. His remains were being sent to Laconner, Washington, on the *Rosalie*. These two men could be the same individual.

Another Malloy who turned up was J.S. Molloy. While this name was elegantly handwritten, there was an ironic tidbit with this appearance. In 1906, a libel suit was filed by Molloy and twenty-four other individuals. (In those days, a libel suit meant a legal action to recover monetary loss.) Two of those twenty-four were "F.M. McGuire" and "Owen McGuire." The owners of the *Clara Nevada* were Harlan P. and W.W. McGuire. Could F.M. and Owen have been related to Harlan P. and W.W.—sons or brothers perhaps? I checked W.W. McGuire's obituary, published in the *Oregonian* on December 19, 1937, and found his son was listed as "Frank L." Later, I found a "P.H. McGuire" mentioned in the Douglas *Island News*, but there was indication that he was in the shipping business. While these are incidentals, they are clearly not matches; the names were close enough to raise this historian's eyebrows.

The only other George Beck I was able to find was in the papers of the territorial governors. On August 23, 1917, a George F(?). Beck wrote from Hoonah to Charles Haukesworth, superintendent of native schools, Bureau of Education, Juneau, about a case of child molesting and whiskey being sold to Indians. The middle initial, however, was hard to read and could have been a "J."

Then there were the passengers. While there was not a single close match with any of the passengers, other than E.W. Saportas, there was a solid match with two men who may very well have been stowaways. William Hemming and George Kasey, whose names were associated with the lifeboat that was discovered by the *Seaolin*, did appear in the historical record. George Kasey appears in the 1900 census as a laborer at the "Juellin Gold Mine" in Berners Bay, a few miles from Comet across the neck of a peninsula. The community of Jualin, the correct spelling, is about ten nautical miles from Eldred Rock.

In an effort to identify Kasey and Hemming, I contacted the newspaper in Rockport, Indiana, the city listed on the letters found in the bundles of clothing in the lifeboat. That letter resulted in correspondence from the youngest granddaughter of William Hemmings—the newspaper had left off the final "s" in the name. She revealed that both Hemmings and Kasey had left their families in Rockport and headed for Alaska in late 1897. Both returned shortly after the turn of the century. Hemmings died on February 10, 1917, in Rockport at the age of fifty-nine. His neighbor and friend, George Kasey, also returned to Rockport and later served as a township trustee for Spencer County, Indiana, for many years. He died in Rockport in 1936. Obviously, both men survived the wreck of the *Clara Nevada*. But were they even onboard? According to most newspaper accounts, "seven Klondikers" were onboard the *Clara Nevada*, but the names were not recorded. Were these two among the Klondikers?

But because the men's belongings appeared in the steamer's lifeboat, Alaskan historian Bob De Armond suggested that these men might have been stowaways. The lifeboats were, after all, a favorite place for men to ride free. Considering that the trip between Skagway and Juneau might have been six or seven hours at the most, the voyage would have been a short but uncomfortable one. That Kasey and Hemmings returned to Rockport broke is a good indication that they may very well have been stowaways. Considering that Kasey and Hemmings survived the wreck, they may very well have been in the only possible place to survive such a catastrophe—a lifeboat.

As to the possibility that Kasey and Hemmings might have simply discovered the abandoned boat and had never been on the *Clara Nevada* at all, it is important to point out that the articles in the *Dyea Trail* and the *Alaska Miner* both stated that the lifeboat had been "stove in and had evidently been on the beach for some time."[36] It would seem unlikely that the men had accidentally discovered the boat and used it for pleasure, wrecked it on their own accord and then left it so close to where flotsam from the wreck of the *Clara Nevada* had accumulated on the shore. The Lynn Canal is hardly the place to be pleasure rowing in February, March or April, particularly for landlubbers from a farming community in America's breadbasket. In those three months respectively, the average temperature is 27.8, 31.2 and 39.0 degrees Fahrenheit.

But there was still the question of a solid match. So far the connections have been tenuous. While Paddy MacDonald, George Kasey and William Hemmings are good candidates for survivors, without rock-solid proof that *anyone* at all survived, any discussion would be purely academic.

But there was one solid match: Captain C.H. Lewis. Between December 18 and December 24, 1899, census taker George P. Morris was traveling up the Yukon River, recording the residents who lived in the area. On the "Yukon River between Circle City & Halfway Island," he registered eight people. One of them was "C.H. Lewis" of Baltimore, Maryland, employed as a "sea captain." The other people on the frozen river listed their occupation as "woodchopper." Living with Lewis was his brother, George G., also of Baltimore.

Is this the same Lewis? Probably. In the 1900 census, there were only five Lewises; none of the others was close to C.H. Lewis by name or occupation. In other nautical records, more than one Lewis is mentioned, but none that are close to C.H. Lewis. The nearest are a Captain S.E. Lewis of the "North Star Line," who appears infrequently in the historical record, and Fred J. Lewis, who was a mariner as well. S.E. Lewis was mentioned once by name in the *Nome Nugget* on August 19, 1903, as being in town transacting business. Fred J. Lewis has been written up frequently but appears to be no relation to C.H. Lewis of the *Clara Nevada*.

Tracing C.H. Lewis through historical documents after the wreck of the *Clara Nevada* proved to be exceedingly difficult. There were some tantalizing historical minutia, but the solid evidence was elusive. There was, for instance, a suit filed by a "G.H. Lewis" in 1905. Earlier than that, in 1901, a suit was filed against a "John Doe Lewis." While this was obviously not a match, what attracted my attention was that the other defendants include "John Doe Barlow" and "John Doe Green." This was tantalizing but by no means definitive. A George Lewis shows up in the *Tanana Directory* in 1907, possibly C.H. Lewis's brother, but, again, there was no solid connection.

Another discovery that raised my expectations was a company doing business in Alaska by the name of Allen & Lewis that had been formed in Multnomah County on November 13, 1897. The company was incorporated to perform a wide variety of functions though primarily to "export and import domestic and foreign merchandise" with regard to wholesale groceries. The company was registered to do business in Alaska, and it began with capital stock for the corporation listed at $400,000, not a paltry sum in those days. While there was no Allen in the initial filing, a C.H. Lewis did join in the firm in November 1907. This C.H. Lewis, however, signed documents as late as June 1942, so it was unlikely that this was the same man.

Later, I discovered that this particular C.H. Lewis was Cicero H. Lewis Jr. and of no relation to the C.H. Lewis that I was researching. A check of signature of Charles H. Lewis clearly showed they were not the same person.[37] I even found a woodcut of Cicero H. Lewis Sr. but nothing of Captain C.H. Lewis. (Cicero Lewis, oddly, was distantly related to George Foster Beck.)

Surprisingly, a check of maritime records revealed nothing either. According to the National Archives, there was no record of a C.H. Lewis in the "List of Officers Licensed" by the Steamship Inspection Service between 1896 and 1902.

If these records are correct, and the census records for 1900 were also correct, what had happened to C.H. Lewis?

Chapter 6

What Happened to C.H. Lewis?

The discovery of a survivor of the *Clara Nevada* disaster—and a significant individual at that—led to two obvious questions: how did C.H. Lewis survive and what happened to the $165,000 in raw gold?

I was haunted by those questions. For two years, I checked archive holdings, government reports, diaries, personal correspondence, business records and court files and sent letters to genealogical societies and maritime museums. I contacted historical archives on both the East and West Coasts. Starting at the Z.J. Loussac Public Library in Anchorage, I read every day of every Alaskan paper from 1900 to 1910. Every source material on the Klondike and Alaska gold rushes was examined. I contacted Canadian sources, as well as the United States State Department, federal records centers, state record centers and private museums. But everywhere I went, I drew a blank. If Lewis had been in Alaska during that time period, I couldn't find him.

Then fate interceded, lending me a hand. In May 1990, the United States government opened a Federal Records Center in Anchorage. Alaskan historians have always had a difficult time doing research on Alaska because the critical historical documents were usually scattered across the United States. Territorial officials were appointed by the president, but when they were replaced, they took their personal papers with them—back to New York, Boston, Atlanta, San Francisco or to wherever else they retired. Business records could be anywhere, depending on the home of record of the parent company.

But finding personal records was easy when compared to state and federal documents. Since Alaska did not become a state until 1959, there was a scrambling of material. Some federal records were left in Alaska where they became de facto state records. Other federal records were scattered and filed on the basis of their subject. Court records could be in Seattle, Juneau, Anchorage or Nome. Indexing was rare.

Researching Alaskan history was made even more difficult by the tremendous cost of travel. The State Library is in Juneau, 600 miles from my home in Anchorage. Seattle and the Federal Records Center are another 600 miles south, and the National Archives are in Washington, D.C., an additional 2,500 miles east. It costs Alaskan historians as much to see the federal records in Seattle as it would a California historian to fly to the National Archives in Washington, D.C.

To allow Alaskans access to their own historical records, it was decided that a small records center for Alaskan-related federal documents would be opened where they would be more accessible to Alaskans. Thus the center was opened—a little less than six miles from my home.

I was the first patron through the door on the first day of the records center's operation. I was quite a surprise to the archive staff. They were still putting on drawer labels as I began pawing through the spools of microfilm, looking for a clue as to the fate of C.H. Lewis. It was historical heaven. Here were all of the primary documents I needed—and on microfilm!

Thus I proceeded through the Coast Guard and Revenue Cutter Service records. Then I went through customs service files, wreck reports, inspection records, sealing inventories, burial lists and government receipts. After that, it was military records, State Department correspondence and miscellaneous files. Roll by roll, I proceeded through the primary documents, hoping—with each rotation of the crank—that C.H. Lewis would have made some mistake, like selling whiskey to the Indians. Or that he had been arrested for a minor offense. Or that he had filled out some federal forms, paid a customs fee or gotten married. What I needed was the name of the ship on which he was a captain. From there, I could find where it was registered and then trace the ownership documents to find out who was a partner on that ship and, perhaps, who had paid for the ship.

But I found nothing. Nada. Niento. Zip. Goose egg. Roll after roll showed me that C.H. Lewis had kept his nose very clean. Then I got a break.

After unending hours of microfilm, on the second-to-last reel on which it was reasonable to expect to find any documentation, the name C.H. Lewis suddenly "popped up." On September 25, 1899, three members of the crew of the *William H. Evans*, registered in Baltimore, Maryland, filed a demand for back pay with the American Consulate in Dawson. The men had left the ship with the permission of the ship's master, "Charles H. Lewis," because "there was a shortage of provisions, and the steamer was on a bar in the Yukon River near Halfway Island between Circle City and Fort Yukon, Alaska." (This was probably in Alaska.) In the documents, Moses Murphy had demanded wages for the previous eighteen months at "$751.67 less $118 and a pair of moccasins [valued at $6.00]." Robert J. Dungan had filed for his wages for eleven months at $687.50. The third seaman, John J. Manton, asked for one year's worth of wages ($1,184.17) plus $200.29, which, he claimed, "I advanced to said Captain Charles H. Lewis...to buy wood for said steamer *William H. Evans*." (Later, after I had discovered the name of the ship, I discovered that at least five crew members had abandoned ship for a rush to Koyukuk country.)

Suddenly the search for the elusive sea captain was over. But this only marked the beginning of yet another journey. Now that I had found him, what would the documentation reveal?

According to the statement of Moses Murphy, he had been hired as a fireman by Lewis for the *Evans* on June 20, 1898. Dungan had been hired on June 1, 1898, and Manton on June 15.

This was odd. According to the *Annual List of Merchant Vessels of the United States*, a United States Department of Commerce publication and the standard source for general information about vessels, the "*W.H. Evans*" was built in 1898 in Ballard, Washington. Yet, the ship was registered in Baltimore, Maryland. Considering that the *Clara Nevada* went down on February 5 and the first crew member that can be documented was hired on June 1, it appeared that Lewis was able to survive a sinking, find passage back to Seattle, buy a steamboat for cash, cross the United States to Baltimore where he registered the steamer, return to the West Coast to outfit the ship, hire a crew and be back in business in a mere 116 days. This seemed hard to believe.

With the name *William H. Evans*, I discovered a court case at the Alaska State Library that resulted in actual documents from the Federal Records Center in Seattle. It was a libel suit filed by the Alaska Packers Association. Though the shipping articles themselves were dated June 20, 1898, the assembled documents revealed a far stranger story.

One of the documents, written by hand, was notes indicating wages due for the time period prior to August 1, 1899, for the purpose of calculating back pay. Captain Lewis was listed as having gone on payroll on "2/9/98."[38] That was only five days after the *Clara Nevada* went down. That was very quick. Even today, the ferry between Haines and Bellingham takes four days.

Most of the crew was listed on board on June 20, 1898. No crew member on the *Evans*, incidentally, matched any name on the *Clara Nevada*.[39]

With the name of the ship, I began rechecking business records and historical sources. From these leads, I was slowly able to piece together the chronicle of C.H. Lewis.

I immediately discovered why I had been unable to find any information on him. He had purchased a new steamboat in early 1898 and run it aground on the Yukon River in August of that same year. The ship had remained high and dry for a year. Then Lewis left Alaska for Baltimore, where he apparently retired. I had been reading Alaskan newspapers between 1900 and 1910, years that Lewis had not been in Alaska. When I began checking documents and newspapers for 1898 and 1899, I did find quite a few references to the *Evans*—none to Lewis, however—and the chronicle I established raised as many questions as it answered.

On September 4, 1897, for instance, well before the wreck of the *Clara Nevada*, Lewis had formed a corporation in West Virginia by the name of the Lewis Klondike Exploration Company. There were five incorporators: Charles H. Lewis, Charles H. Hopkins, James H. Preston, William F. Rogers and Charles W. Lewis, all of whom were from Baltimore. The corporation was capitalized at $399,964, with C.H. Lewis as the largest shareholder with 75 percent of the stock. The corporation had been established "for the purpose of manufacturing, building, equipping, fitting, furnishing, chartering, operating, leasing or owning steam, sail or other boats," as well as a wide variety of gold-seeking and mineral transport activities.[40] This raised an important question. If Lewis, personally, had $300,000 *cash* to drop into a corporation to run steamboats, what was he doing on the *Clara Nevada*, a penny-ante tramp steamer? (There is also the possibility that the firm had no money, and that the figure of $399,964 was simply snatched out of thin air for the purposes of forming a corporation and establishing a par value of stock. If this was the case, the choice of $399,964 is an odd choice. More than likely, the dollars are accurate, for it would be close to impossible to form a corporation to do what was proposed with no money.)

Perhaps the answer was in the history of the shipping industry at that moment in time. For the maritime industry, the Klondike gold rush came at exactly the right moment. The Pacific Northwest had been experiencing an economic downturn, which reached its lowest ebb in 1897. Maritime giants were crashing like ancient trees in driving storms, and East Coast firms were moving aggressively into the maritime trade vacuum of the international market. Just as it appeared that the West Coast sea trade was issuing its death rattle, the *Excelsior* landed, and the industry rose like a phoenix from the ashes of its own pyre. Investment dollars poured into Seattle, and within a year, the Pacific Northwest maritime industry was once again a robust beast.

But the gold rush traffic to Dyea and Skagway was not the only route to the Klondike fields. There was another American sea passage to the Klondike gold fields, and it had its advantages: it was known as the "Rich Man's Route" because of the cost of travel.

The route started in Seattle but then ran across the North Pacific to Unalaska. The ships would pass through Unimak Pass in the Aleutian chain of islands and into the Bering Sea. They would usually stop at Unalaska, refuel and then steam north to St. Michael and the mouth of the Yukon River. There the passengers would transfer to shallow-bottom river craft and be transported up two thousand miles of Yukon River to Dawson and the gold fields of the Klondike.

However, because the trip was so expensive, many stampeders opted to charter vessels rather than paying for passage and cargo space. With ships renting for $400 a day—and trips lasting as long as ten months if the vessel became frozen in the Yukon River for the winter—many of the Argonauts formed exploration companies to reduce their costs. Usually underfunded, these companies often came apart as fast as they were put together.

Shipping companies were making fortunes on this route, and Lewis undoubtedly realized that this was a market worth tapping. After all, the longer the sea voyage, the more he would make, and a ten-month voyage at $400 a day meant a great deal of profit.

But even with the initial investment, the *Evans* proved to be too expensive. As a result, more partners had to be added to the Lewis Exploration Company, including a man by the name of William H. Evans, presumably after whom the vessel was named. The company also changed its name to the Evans Klondike Exploration Company. More money was still needed, so twenty-seven smaller investors were added.

The acquisition by Evans must have been fairly soon after the arrival of C.H. Lewis in Seattle since the Master-Carpenter's Certificate for the completion of the steamer *W.H. Evans* for the "Lewis-Klondike Expedition Co." was issued on April 11, 1898.

As an ironic aside, the acquisition of the *Evans* was exactly two weeks before the wreck report for the *Clara Nevada* was officially filed. Since the *Clara Nevada* was registered on January 26 and its certificate was turned in on March 31, it may well have had the shortest life span of any ship in the gold rush flotilla.[41]

Even with the influx of money from William H. Evans, there was not enough to complete work on the *Evans*. It left the Ballard shipyards with interior work still to be done. That work was contracted locally but was made exceedingly difficult because many of the lesser contributors to the expanded exploration company were now living onboard, as they had no money to reside elsewhere.

Lewis and his partners apparently figured on making a killing on the steamboat trade. Toward the end of March 1898, he began advertising daily in the *Post-Intelligencer*. (I could find no such advertisements in the *Seattle Times*.) "The Lewis-Klondike Expedition Company," the announcement stated—even though the official name of the defunct business was in error—"will dispatch the FIRST CLASS ENTIRELY NEW STEAMER *W. H. Evans* For All Points on Yukon River." The advertisement further stated that the scheduled date of departure was "about April 25th," with the *Evans* being in the tow of "Steamer *Rival* or *Protection*," with connections at the mouth of the Yukon with *Steamship City of Seattle*. This display advertisement ran unchanged through April 22, three days before the *Evans* was supposed to depart. It did not, in fact, depart until two months later.

But the refitting period was not without incident. On May 25, just as the steamer was about to leave, it was hit with a libel suit. Two litigants, the Washington National Bank of Seattle and C.R. Sanborn, an equipment repair mechanic, asked the court to hold the *Evans* until $1,262.59 could be paid. The Bank of Seattle wanted $500 it had given to Lewis to pay for wages, and Sanborn wanted his pay for equipment repair. Lewis must have been able to raise the money quickly for the *Evans*, as no court order held her in port.

Ironically, while Lewis was fretting over the libel against the *Evans*, his former ship, the *Eugene*, was being sued. R.J. Young and W.D. Wallan, two former passengers, were suing the Willamette & Columbia River Towing Company and the Yukon Transportation Company for their money back.[42] Both suits appeared in the same paper on the same page-one column.

Finally, on June 20, the *Evans* departed with the largest fleet ever to have left Seattle for the Klondike.[43] Four ocean-going steamers loaded with cargo and passengers left towing twelve river steamers "identical from stem to stern in equipment, machinery and decking." The *Evans* and *The Northern Lights* were being towed by the *Noyo*. Due to some last-minute changes in the towing arrangements, the *Evans* was eventually paired with the *Alfred J. Beach*.

It must have been a strange-looking flotilla, even for the gold rush traffic. The ships in tow "had been planked up, snow-plow fashion" to prevent waves from washing [their] main deck[s]" and had sack coal on deck for the journey. The trip was expected to take twenty-five days, with the convoy hugging the shoreline of Canada and Alaska rather than risking the open waters of the north Pacific and Bering Sea.[44]

The fact that the *Evans* was being towed also opened a loophole for an extra bit of profit. Government inspectors would not allow passengers on a river steamer to go across the open ocean, but since the river steamer was being pulled by an ocean steamship, it was perfectly legal to carry passengers. It was very dangerous, but it was still, nevertheless, legal.

There were many that were appalled at this example of greed on behalf of the steamship companies. Even the *Post-Intelligencer*, which was not known for its enthusiasm in dissuading passengers from using Seattle-based steamers, noted the danger in this practice. Editorially, in a news article, the paper suggested that the passengers on the *Evans* were surely those "whose anxiety to reach the land of gold has overwhelmed their judgment and caused them to take the desperate chances incident to such a voyage."[45]

Unfortunately, the ship did not leave its bad luck on the dock in Seattle. The *Evans* slipped her tow lines three times between Seattle and Port Angeles. Then, shortly after leaving Port Angeles, she was damaged at the mouth of the Straits of Juan de Fuca and requested to be towed back to the dock. The master of the tow ship, the *Noyo*, Captain Hanson, refused to return the *Evans* to port. This, Lewis later claimed in a suit, was because the officers of the *Noyo*, "in collusion with the officers and owners of the *Beach*," intended to run and therefore damage the *Evans* so badly it would break up, thus "lightning the load of the *Noyo* and hasten her arrival at St. Michael while at the same time nominally performing her contract."

Bad weather, however, forced the *Noyo* to seek shelter herself at Port Angeles. But the imbroglio was far from over. Once ashore, the master of the *Noyo* attempted to have the *Evans* declared unseaworthy. This failed, and Lewis's complaints resulted in the removal of Captain Hanson. Unfortunately, Hanson's replacement, Captain Edgett, turned out to be just as disreputable.[46]

Long before the *Evans* even arrived in Alaskan waters, disaster struck again.[47] On July 21, the *Post-Intelligencer* reported that the *Evans* and four other river steamers had gone to the bottom. Later, a more accurate picture was reported. The *Noyo* had been pulling two river steamers, the *Evans* and the *Alfred J. Beach*, when a violent storm came up the morning of July 2. The *Beach*, the last ship in tow, took on water in seas in which the waves were so high that they were cresting over the snowplow arrangement on the stern, "sweeping her main deck and carrying off portions of the light deck house."

Unfortunately, the *Beach* was so poorly built that it was surprising she made it as far as she did. Five hundred miles off shore, her "hog chains parted" and her "boiler went through down through the bottom of the ship." (The *Beach* was not the only ship in such poor condition. Of the thirty-two ships that left Seattle in the great flotilla, only eight made St. Michael. They averaged four knots, hardly a speed to inspire confidence in passengers and had "troubles of all kinds."[48])

By 3:00 a.m., it was apparent that the *Beach* was going under, so the passengers abandoned ship and boarded the *Evans*. Several hours later, the *Evans* showed signs of breaking up. The *Noyo* then turned and ran toward shore where she beached the badly leaking *Evans* at the Tlingit village of Howkan, south of Wrangell, on the Fourth of July.

The sad state of affairs was still far from over. When the passengers from the two wrecks boarded the *Noyo*, they discovered that the steamer's deck was so stacked with cargo that the vessel could easily turn turtle at any moment. Worse, the hatches had not been closed upon leaving Seattle, and in the case of a storm such as the one they had just encountered, the ship might very well fill with water and sink. According to the *Post-Intelligencer* on July 22, the passengers rioted and began throwing freight overboard until an armed officer of the *Noyo* "put a stop to the trouble."

Now Lewis wanted vengeance for being left beached. He probably also wanted back the $40,000 he had paid to get the *Evans* to St. Michael. While the *Evans* was being repaired, Lewis returned to Seattle. There, on August 20, he and his company filed a "sensation libel for heavy damages" against the *Noyo*, charging the master, H.W. Edgett, with "having deliberately attempting to wreck the *Evans* in order to nominally perform the *Noyo*'s towing contract without in fact doing so."

In the suit for $24,000 in damages, Lewis claimed that the *Noyo* was contracted to take the *Evans* along a route that hugged the coastline. The master of the *Noyo*, however, "in collusion with her officers with the officers and owners of the *Alfred J. Beach*," attempted to take both steamers "to the open ocean and directly across the North Pacific Ocean to Unimak Pass…and thence to St. Michael."[49]

While Lewis must have been furious, it is ironic that the master of the *Evans* was suing the *Noyo* in August 1898 for an action that was almost identical to that of which he had performed just a year earlier with the *Eugene*. Perhaps Lewis had learned his lesson. Or perhaps he had no choice but to sue for damages to repair to the *Evans*, which had been badly damaged by the foolhardy actions of the master of the *Noyo*.

Further, in the suit, Lewis stated that he had protested the change in direction of travel but had been ignored. When the *Noyo* was about one hundred miles offshore, the *Evans* "signaled that she was in great danger." The response of the skipper of the *Noyo* was to demand that Lewis "sign a release of the *Noyo* from all liabilities from her contract and pay compensation for towing her to a safe anchorage." When Lewis refused to sign, the captain of the *Noyo* stated he would "tow the *W.H. Evans* to sea until she sank."

Four hours later, the *Noyo* had no choice but to turn back because of heavy weather. Lewis was approached again to sign the release, and again, he refused. Then the *Noyo* cut the *Evans* loose and refused to tow her any further. The situation was complicated by the sinking of the *Alfred J. Beach*. Unburdened with any steamers to tow, the *Noyo* arrived, alone, in St. Michael on July 16.

Back in Seattle, considering the *Evans/Noyo* story to be one of those strange-but-true episodes in the saga of the steamboat frontier, the *Post-Intelligencer* listed it that way. "Strange Tale of the Seas," the headline blared. The paper was so convinced that Lewis was telling the truth, it also ran the follow-up line as "*Noyo*'s Captain Charged with Deep-Dyed Villainy."

Though the *Evans* was beached, Lewis swore that he would repair the vessel and head north again by himself. He and his crew "rigged her with a stern wheel" and repaired the ship so that they could be northbound by July 5. When the steamer had been repaired, Lewis set out for the mouth of the Yukon River some two thousand miles away, alone, in some of the most treacherous waters of the world.

But the ship's bad luck continued to plague her. Because she was forced to hug the coastline, the trip was much longer than expected. As a result, coal was in very short supply. How Lewis expected to fuel the ship is unclear. If he figured that he could use wood, he was in for a surprise. To this day, there is a lamentable but true expression that there is "a woman behind every tree" in the Aleutians. (The Aleutians are famous for their dearth of both women and trees.) But, as the steamer chose its route close to shore, she stumbled onto two coaling locations without foreknowledge of their existence.

On July 27, 1898, the *Evans* took on twenty-five tons of coal at Chignik on the southern shore of the Aleutians and wrote the Alaska Packers Association a check for $750, probably knowing full well that the check would bounce. On August 11, 1898, the steamer took on an additional twenty-five tons of coal from the Alaska Packers Association, this time in Nushagak on the northern side of the Aleutians. Once again, the coal was purchased with a rubber check.

According to the *Post-Intelligencer*, the *Evans* reached St. Michael on August 30. Here they took on a load for the Alaska Exploration Company and proceeded toward the Yukon River and Dawson. This fit with the data recorded by the census taker, as both C.H. Lewis and his brother claimed they had entered Alaska in August 1898. Then the *Evans* headed for the Yukon River, trying to make it as far into the watershed as possible before it was frozen in for the winter.

Bad luck, however, still followed C.H. Lewis and the *Evans*. Lewis steamed by the mouth of the Yukon River without recognizing it. Mistaking a landmark, he headed up a shallow arm of the mighty river and ended up on a silt bar. The steamship *Louise*, having made a similar mistake of her own, was coming down Kwikpak Pass and saw the *Evans* in trouble. The *Louise* backed away before she grounded herself as well. Fortunately for both captains, they were not too deep into the Yukon Delta. With the next tide, they both rose with the tide out of the silt. The *Evans* backed out to sea to find the correct entrance to the Yukon River.[50]

The Yukon, however, was not a kind river to steamer traffic. "It's the dirtiest river in the world," Captain Peter M. Anderson, a seasoned Yukon River steamboat pilot, was quoted in an interview with the *Post-Intelligencer*. "You never know where you are at. There is not enough water in the first place, and the sand bars seem to move around in your way. The riverboats go aground, break their [two] chains and get terribly out of line."

Going aground was a constant danger. If the steamboat went aground going upstream, the cargo was sold at a great loss. This lightened the vessel and often allowed the steamer to float free of the river bottom. But if the steamer was empty and headed downstream when it struck the sandbar, it was considered lost forever. It would have no cargo to unload, and unless it could float free before the ice came, it was doomed. By "May of the next year," noted Captain H.L. Adams, "the ice run and breakup would [have demolished] the boat, and there would not [have been] a plank left."

Though the *Evans* was able to make it to the actual mouth of the Yukon, her luck did not change once she entered the watershed. She made it as far as the Dall River and spent the winter of 1898 there. The next spring, she proceeded upriver. Unfortunately she lacked the power to proceed, and at Circle, the *Evans* turned its cargo over to another steamer, then turned around and started back downriver.

But now the weather conspired against her. There was record water level that year. In June, the Yukon was twenty-four feet above low water. In other places, the water was reported to be sixty-two feet above the low water line. In Dawson, the streets were flooded, and passage from one side of town to the other required a boat.[51]

As the water receded, it created problems for the river steamers. By mid-July, Fred Seigel of Mount Vernon, on the *Hattie I. Phelps*, reported that the water level was so low that "we had two pilots and four *siwashes* and we were constantly bumping sand bars. The bars are constantly changing and navigation is therefore very hard." Seigel also reported that there were a number of steamboats stuck in the river, including the *Seattle No. l*, which was "hard and fast on a bar with poor hope of getting off," as well as the *Weare* and *Bella*, which had been "pushed on the banks by ice."

In August 1898, the Yukon was so low that it was reported that few boats were making it into Dawson. For reasons we will never know, thirty miles south of Circle, the *Evans* finished her maiden voyage stuck high and dry. When the ice went out the next year, she was stuck where she could never move. Thus, when the *William H. Evans* became grounded near Halfway Island, it was at least with company along the river.

With the *Evans* on the sand bar, the end was slow in coming. Lewis was able to stave off destruction during breakup in 1900 by filling the *Evans*'s hold with water. This put an estimated 1,200 pounds of weight in the ship, giving the hull the strength to withstand the "immense force of ice as it piled up on her at break up."

Yet another irony of the *Evans* was brought to light by the *Seattle Times* in February 1899. To profit from every cubic foot of cargo space, the *Evans* took on pass-through cargo—according to Admiralty court records, 162½ tons of it—for Dawson rather than food for its crew. Then, when the ship became ice locked in the river, Lewis had to buy food at inflated prices to feed himself and his crew.

The crew and passengers apparently remained aboard through the winter and spring, for it was not until the next year that the crew member who sued for back wages left the *William H. Evans*. By then, July 30, 1899, the steamer had been on the sandbar for almost a year. Prospector Walter R. Curtin saw the ship aground on June 19 and noted that it was "seemingly out of the water. She looked as though she were going to be on dry land all summer." In the appendix to his book, *Yukon Passage*, Curtin further noted, saying, "[The *Evans*] ran aground in the Yukon Flats in her first year; and the falling water left her stranded so high that I do not think she ever got off, at least not while there was any business for her."

As far as the historical record goes, Curtin was apparently correct. The *Evans* did not sail again that year. That December, the census taker found Lewis and the *Evans* iced in on the same sand bar.

Interestingly, according to Captain H.L. Adams, maritime historian, there is a bar on the Yukon River between Fort Yukon and Circle directly across the river from Halfway Island named Evans Bar. This name originated on a chart by Captain Charles Y. Mahnquist, who drew the chart in 1898. It would probably be safe to assume that this bar was named after the *William H. Evans*.

Being iced in during the winter and grounded during the summer were not Lewis's only problems. In addition to his crew abandoning the steamer, he was also being sued. When Dungan, one of the crew members suing for back wages, was in the office of the United States consul in Dawson on November 2, 1899, he handed the consul a document that he claimed he had found in the engine room of the steamer. It was a notice that the Alaska Packers Association was suing the *Evans* for supplies furnished. Lewis's bad credit had finally caught up with him.

When neither check to the Alaska Packers Association cleared—and after the *Evans* was finally located almost a year later—the packers filed a libel suit against the steamship on March 22, 1899. The vessel was seized on August 5, which was not hard to do since it had gone aground. A court date was set for December 1, 1899. When the vessel could be floated, it was ordered to proceed to Skagway, where it was to be sold at auction. Wages totaling $13,494.66 were ordered recovered for the crew. The notice was authorized in Circle City on August 5, 1899, by United States marshal James M. Shoup.

But Lewis was not through with the *Evans* yet. Even though he and his crew claimed on August 1, 1899, that "each and every intervening libelant is a poor person and without means to pay the fees and costs of this suit," Lewis was able to buy back his own ship. After all, who in Skagway would want to buy a steamer that was stranded on the Yukon River a thousand miles away? On June 23, 1900, when the ship finally went for auction, Lewis, "being the highest and best bidder and the only bidder," bought the *Evans* in Skagway for $13,000. That sum would have paid off the crew but apparently not the Alaska packers.

Even after Lewis bought his own ship back, there were problems. The Alaska Exploration Company filed a libel suit against the *William H. Evans* in August 1899. The suit was still pending in September of the next year when the United States government formed the Second Judicial District and transferred the suit to Nome. Nothing came of the suit, most likely because the *William H. Evans* no longer existed when the case finally went to court.

In the end, the *William H. Evans* could not be moved. Fearful that breakup the next spring, 1901, would damage the ship too greatly to be repaired, the vessel was sold again. This time it was scrapped. What could be salvaged was loaded onto a barge and dragged into Dawson by the Alaska Commercial Company's steamer *Bella*. There the machinery was bought and used by the Yukon Saw Mill.

With the sale of the *Evans*, Lewis again disappeared from the historical record. This time, however, I had a handful of leads. As I unraveled Lewis's life, it became clear that he was as much the epitome of the rough-and-tumble mariner at the end of the nineteenth century as one could find.

Lewis was apparently a born adventurer, perhaps the last of his breed. According to his obituary, he was born in 1834 and was one of the first men to go to the California gold fields by sea, at age fifteen, most likely as a cabin boy. Then, in 1853, as a teenager, he went with the Perry Expedition to Japan. During his twenties, he smuggled guns and ammunition to the Confederacy through the Union naval blockade. Later, ironically, he was "connected with the United States Geodetic Survey" in the surveying of the Gulf of Mexico, the same governmental body that owned the *Hassler*, which became the *Clara Nevada*.

Lewis had deep roots in Baltimore. I discovered that even though he was a sea captain on the West Coast, he maintained a home in Baltimore. The *Baltimore City Directory* for 1888 lists C.H. Lewis, mariner, as living on 1809 Light Street. A George C. Lewis, presumably his brother, was residing at 219 East Hamburg Street in the same city. They were both working as masters on different coasts. C.H. was listed as a first officer on the *Idaho* in Pacific Northwest waters that year, while G.C. Lewis appears in the historical record as master of the *Alsenborn*, enrolled in Baltimore on September 18, 1888.[52]

I also discovered that the information on the wife of C.H. Lewis, as presented in the Seattle press, was in error. C.H. Lewis had a wife, but there are few West Coast references to her. Two were from the *Post-Intelligencer*, two days apart, and were contradictory. On February 18, she was reported to be in Portland and "very ill." Two days letter, the same paper placed her in Peru. Supposedly she had accompanied Lewis to Peru when the *George W. Elder* went south "during the Chilean-Peruvian War." Though Lewis returned to the United States, his wife did not, as "she found the associations of Peru better to her liking than a trip up the coast." Considering that her husband was the captain of the steamship at the time, it could have been insinuated that her staying in Peru was tantamount to a divorce.

However, when C.H. Lewis was finally located in the 1900 census in Baltimore, another tidbit in his life came to light. He did have a wife, and in 1900, she was very much alive and not in Peru. Her name was Sara F. Lewis, she was nine years younger than her husband and the couple was living with their son, two daughters, a son-in-law and a grandson. The couple had been married for thirty-five years in 1900 and had produced eight children. The remaining son at home, Robert J. Lewis, age twenty-six, was also listed as a "mariner." The entire brood lived at 1429 William Street, in South Baltimore near Federal Hill.[53]

However, late in my investigation, I made contact with the great-great-granddaughter of Lewis. She confirmed that while Lewis was not living in San Francisco in 1900, his family was. The 1900 census lists a large family as "Household," with a selection of last names as follows (with their age in parenthesis): Mary Lewis (thirty-nine), Alvin Fouratt (sixteen), Rita Fouratt (fourteen), Charles Lewis (seven), Honora Lewis (three), Grace I. Lewis (fourteen), Victor Street (thirty-five), Samuel Uncovich (twenty-two) and Spiro Radovich (twenty-eight). I was told the two Fouratts were from a previous marriage. Because of the ages of Street, Uncovich and Radovich, I'm inclined to believe they were boarders. And this collection did not include those in Baltimore in the same census. According to census records, the Lewis family was living at 25 Tenth Avenue. The actual house in which they were living was apparently dynamited in 1906 to create a firebreak during the great earthquake and fire. The census in 1900 in San Francisco lists Grace Lewis, the wife, as a widow, so apparently she thought her husband was dead.

The presence of C.H. Lewis in the United States census in June 1900 in Baltimore raised a nagging question: how could C.H. Lewis have been in both Alaska and Maryland at the same time? Was he in two places at the same time? Actually, he wasn't. On December 6, 1899, he was in Alaska. On June 7, 1900, he was in Baltimore. He was counted by the census twice. Considering that he was back in Skagway on June 21, two weeks later, he was apparently in Baltimore to raise the $13,000 necessary to buy back the *Evans*.

The deeper I dug, the harder it was to believe that Lewis could have been given command of a ship. Perhaps the only reason was that the gold rush required all types of mariners, regardless of talent or record. When it came to his record, Lewis certainly had one that did not bode well for his employers.

Piece by piece, I put Lewis's life on paper. Possibly the first boat he skippered was a schooner by the name of *Pilot's Bride*. But Lewis did not command it long. It became stranded on Nestucca Bar, Oregon, and was declared a complete loss in August 1881.[54]

Thereafter, his record did not get any better. On December 21, 1887, Lewis was serving as the captain of the *San Vincente*. Forty-five miles south of San Francisco, the ship caught fire from an unknown source and burned to the water line; then it sank. Some of the crew managed to get off in a lifeboat, which was subsequently lost at sea.

Six crew members in another lifeboat were successful in escaping the burning ship and were picked up by a passing vessel. They were lucky to have escaped when they did. By the time they were rescued, it was reported that the flames from the burning *San Vincente* were so high that they appeared as "spires of flame leap[ing] up in the darkness, illuminating the waters for miles around." The cargo was listed as "general merchandise," and the ship was uninsured at the time.

Giving credit where credit is due, Lewis apparently stayed on deck until the last possible moment, looking for eleven missing crewmen. According to one account, he stayed on deck so long that he had to escape on a rapidly made raft because both lifeboats were too far away to be reached by swimming. Lewis was also burned badly in the fire and probably carried the scars of that incident for the rest of his life.

The ship that rescued the men was the *Queen*, the same ship which was to bring the news of the sinking of the *Clara Nevada* to Seattle eleven years later. Another odd bit of synchronicity, the fire was attributed to "the explosion of a can of oil, which was on top of the boiler."[55]

Lewis later served on the *Alexander Duncan*, which was captained by a man who would cross his path many times in the future: James Carroll. But he could not have served on the ship very long, for it was wrecked on Fort Point in September 1885.[56] Lewis also served on the *Salinas*, which sank in the "heaviest gale known for years" in Monterey Bay.[57] Whether Lewis was onboard when these two ships went down is not recorded.

According to the Marine Engineers Beneficial Association, Lewis was listed as a veteran of twenty years' service with the ORNC (Oregon Railroad & Navigation Company) and the Pacific Coast Steamship Company, the forerunner of the Admiral Line. During that time, the beneficial association stated, he had served or commanded on a variety of ships, including the *Idaho* and the *Michigan*. (Strangely, Cicero Hunt Lewis, the "other" C.H. Lewis, was an incorporator of the ORNC.[58])

With regard to the *Idaho*, on which C.H. Lewis was credited with service, it was seized in 1886 by the U.S. Revenue cutter *Oliver Wolcott* at the direction of H.F. Beecher, the customs inspector at Port Townsend. Upon examining the cargo, Beecher discovered that the *Idaho* was carrying $48,000 worth of opium. Customs officials raided a fish cannery in Alaska owned by the *Idaho*'s master, where they found an additional eleven barrels of opium. Released, the *Idaho* was seized again, three months later, with another six hundred pounds of opium. The ship was placed under $30,000 bond, but while the case was being appealed, it was wrecked on November 29, 1889. When customs officials examined what was left of the ship shortly thereafter, they discovered several hundred more pounds of opium onboard!

In all fairness, however, it should be stated that opium smuggling was not the crime then that it is now. Anyone could buy and sell the substance. But there was a heavy duty on it. Opium sold for sixteen dollars a pound, but it had a duty of ten dollars a pound. The legal case involving the *Idaho* was, thus, not about the legality of opium but on the avoidance of duty on that import.[59]

Interestingly, the final judgment against the *Idaho* was nominal. In fact, in spite of the confiscated evidence and the obvious proof of ongoing criminal activity, the fine was so low that it did not even compensate the United States government for the $8,000 in fees it had paid to informers. While records do not indicate Lewis's position on the ship at the time of the initial federal investigation, in September 1888, he was listed as "first officer," a command position.[60]

Some newspapers credit Lewis with serving as the captain of the *Michigan*, but the historical records are lacking. While Gibbs and McCurdy stated that Lewis was the master of the ship, the *Daily Astorian* refers to the captain of the vessel in its columns as "S.F. Graves," though the incident is reminiscent of Lewis's headstrong antics.[61]

In November 1890, while en route from Puget Sound to Portland, the *Michigan* caught fire at sea. The captain, Lewis or Graves, fastened down the hatches to smother the fire and made a "wild run" down the coast until the ship reached Astoria. Dockside, the Astoria Fire Department filled her with water as hundreds of spectators lined the docks. "It was veritable volcano," the *Daily Astorian* reported, "confined among 500 barrels of lime and 2,300 cases of salmon, with a lot of coal on one side to help it out." It took seven hours to quench the flames and left "nine or ten feet of water in the hold." Repaired, the *Michigan* set out and was shortly thereafter seized by customs for smuggling and put under heavy bond.

The disasters of the *Eugene*, *Clara Nevada* and the *William H. Evans* came thereafter. But this did not end Lewis's maritime career. Even though he was in his sixties when he returned to Baltimore after the beaching of the *William*

H. Evans, he continued to work in the shipping industry. In the first decade of the twentieth century, Lewis was connected with yet another shipping firm, the Baltimore and Carolina Steamship Company, which offered passenger and freight service between Baltimore and Miami, a four-day trip. The company had four ships, *Esther Weems*, *Mary Weems*, *Nancy Weems* and *Georgianna Weems*—named, perhaps, after the relatives of the president of the company, Mason L. Weems Williams.

The Baltimore and Carolina Steamship Company was founded in 1905, and Lewis probably worked for the lines as an investor, as he would have been sixty-six when the company was formed, a bit late in life to be a skipper. The last historical reference I could find on the Baltimore and Carolina Steamship Company was in 1923, when the Maritime Index reported that the company raised money through a bond sale.

Lewis died June 7, 1917, and his obituary ran the next day in the *Baltimore Sun*. Lewis was survived by his wife, Sarah F. Lewis, and four children: Mrs. Mary A. Forsythe, Miss Sadie F. Lewis, Captain Charles F. Lewis and Captain Robert J. Lewis. Though I sent letters to every Lewis and Forsythe listed in the 1988 Baltimore phone book, I did not receive a single positive response.

Chapter 7

The End of the Maritime Frontier

Perhaps the ultimate irony of C.H. Lewis's life was the beaching of the *William H. Evans.* Here he was, at the top of his profession in his peak earning years, the captain and owner of his own vessel, midstream in one of the greatest gold rushes in the history of mankind, a stampede that depended primarily upon water transportation—and fate had left him sitting high and dry while a flood of steamships passed him going upriver. No doubt he realized that every morning he didn't travel, he was losing $500 in income—$14,000 today.

But it was also ironic. In the decades between 1880 and 1900, he had skippered five ships, the *Pilot's Bride*, *San Vincente*, *Eugene*, *Clara Nevada* and the *William H. Evans*. Of these, two had been beached, two burned and one sunk. Only the *Eugene* remained above water when Lewis's tenure as captain was through, but it had bankrupted its owners and come close to turning turtle as well. Examining his record, one would be hard pressed to call Lewis a competent skipper or successful businessman.

Being responsible for that many ships that met disaster hardly looked good on one's record. Other captains who had done far less were punished far more severely than Lewis ever was. In February 1899, for instance, Captain Jessen of the *Homer* had his license suspended for thirty days. Jessen's transgression was backing into the steamer *Al-Ki* at a very low speed. Total damage was about $3,000. This was not even pocket change compared to the damage of the *Clara Nevada*, not to mention loss of life.

But times were changing. Though Lewis may not have known it at the time, the *William H. Evans* and the rest of the gold rush fleet was doomed. The end of their era was rapidly approaching in two forms.

The first was a railroad: the White Pass & Yukon Railroad, "the road that couldn't be built." An organizational hybrid, it was actually three railroads, one incorporated in the United States, another in British Columbia and a third in the Yukon. The first passengers went over White Pass in February 1899, and the first through passengers to Lake Bennet made it in July of that year. By the time Lewis resolved his legal difficulties, the Yukon River route was defunct. It was cheaper to go by steamer to Skagway and then rail. As of August 1899, trains ran regularly between Skagway and White Horse.

The second blow was the Nome gold rush. On May 6, 1899, the *Post-Intelligencer* ran a small filler, telling "of a marvelously rich gold discovery at Cape Nome, halfway between Golovin Bay and Bering Strait, near the entrance of Norton Sound." Suddenly the attention of the world was focused on the golden beaches of Nome on the Bering Sea. Overnight, the Klondike in Canada's Interior lost its luster. In October, the *Post-Intelligencer* reported that Dawson was emptying, as much of its population joined in the rush to Nome. So many were leaving, reported some stampeders, that "men are getting scarce in the Klondike metropolis." Lewis, still beached, had to suffer the indignity of seeing the rush to Dawson reverse itself on the river.

The most significant nails in the coffin of the Yukon River steamboat businessman were the twin forces of competition and regulation. A shiver must have whisked down the back of every gypsy steamboat operator when it was announced that Alaska was to have a body of codified laws. In June 1899, Congressman Vespasian Warner of Clinton, Illinois, went north to see firsthand what was going on in Alaska.

Warner was concerned that Alaska was still a wide-open frontier. Codified law, he believed, would civilize the territory, and a civilized Alaska would attract investment capital. But before the investment capital could be attracted to the northland, there had to be civil assurances that investments would be safe. Alaska's banks had assets "up to the neighborhood of $7,000,000," Warner told the *Post-Intelligencer*. With that kind of a capital base, "Eastern money is not now asking as to the stability of [Alaskan] business and growth but only the character of the particular investment for which it is wanted." In other words, Warner felt Alaska had enough cash to interest the eastern banking establishment.

But it wasn't only eastern investors who were concerned about the fiscal integrity of Alaska, Alaskans wondered about it. In August 1899, just ahead of freeze up, the newly elected congressman from Washington State, W.L. Jones, fresh from a trip to Alaska, gave an exclusive interview to the *Post-*

Intelligencer. He stated that the Alaskans considered themselves disenfranchised Americans. They wanted more courts, a just tax system and more equitable distribution of federal dollars.

Alaskans began taking a greater part in shaping their own destiny. As Alaska's population grew, Alaskans began to demand their federal rights. On August 15, 1899, a contingent of Alaskans visited Juneau to call a territory-wide meeting in October to protest the "outrageous neglect and ignorance manifested by our national legislative body." They then proceeded to decry their lack of representation in Washington, D.C., and declared what was akin to a declaration of independence for Alaskans.

Virtually absent earlier, law and order were coming to the northland. In 1896, Alaska's law consisted of a single judge, marshal and ten deputies, along with a handful of quasi-law-enforcement employees, such as Indian policemen under the Department of the Interior. Justice had generally been the function of miners' councils. The community elected representatives who dealt out punishment from hanging to whipping. But by 1897, Congress was stepping to the fore. Army personnel were sent to reconnoiter the location of forts. Two were eventually established at Eagle and Tanana. A Revenue cutter, the *Nunivak*, was assigned to patrol the Yukon River.

With competition came regulation. In May 1898, twenty-six carloads of boats arrived in Everett for shipment to the Yukon. In July 1898, the Seattle inspector of hulls and boilers stated that Yukon River steamers would no longer be allowed carry passengers from San Francisco to St. Michael. There was also the formation of an Alaska Traffic Association to stop rate cutting and stabilize the prices in the steamboat trade. They established a base ticket price of $300 for first class and $250 for steerage from Seattle to Dawson. Steamships would be classified as to condition and accommodations, and then a price scale would be assigned.

Predictably, this last scheme was a grand idea, but the coalition couldn't last. There was still too much money to be made on the northern route. The association lasted all of two weeks, and then it fell apart, as its membership went right ahead and cut rates anyway.

Perhaps just as critical to Lewis, winter traffic in 1898 began picking up by the end of the year. Worse, while the Yukon was not navigable during the winter, the Inside Passage was packed with ships on their way to Dyea and Skagway. "Travel to Alaska has set in with greater magnitude than has ever been known at this time of year," the *Post-Intelligencer* noted in December 1898. Then, in the article immediately beneath that, it was announced that the Chilkoot Pass trams had been put into good condition and would begin operating again on the first day of January at the rate of six and a half cents a pound. By comparison, in early 1898, Tlingit were charging one dollar a pound to pack the same food and supplies.

New modes of transportation began to emerge because long-term investment in northern transportation was still deemed profitable. On February 10, 1899, the *Post-Intelligencer* ran a sidebar on the movement of the ship-building industry from the East Coast to Seattle, particularly with regard to wooden sailing vessels. Seattle, the *Post-Intelligencer* revealed, was a natural port for such an industry. First, there was the ability to build a superior craft cheaper with better material. Better material was Washington fir, which could stand "bumping and strains better" because of its "rubber-like quality" that kept it from "cracking when driven against piers or docks." Just as important was the opening of China to American wheat. Though this was in the "years to come," the *Post-Intelligencer* assured its readers that "Eastern men realize [the] great future of the [Puget] Sound [and] they will come slowly at first, and then in droves, to take advantage of its great superiority."

This might not have been so farfetched. In fact, in July 1899, the United States government stated that fiscal 1899 had been the best year in a decade for shipbuilding. Excluding 1891, it had been the best year in the past quarter century. More than 1,400 vessels had been constructed in the previous twelve months compared to more than 1,500 made by Great Britain. "Nearly all our tonnage is built to navigate in the coasting trade reserved to American vessels," the report stated.

For the Yukon River traffic, shipbuilders were experimenting with a new model, a wide-bodied, shallow-draft steamer for river travel. Unveiled in April 1899, this new breed of river craft was powered by six engines, one for each propeller. It was the only such vessel in the Western Hemisphere, and only two others were in existence, both on the Nile. The track record of this craft, so to speak, was impressive and precisely suited for travel in waterways like the Yukon. The craft proved to be a "wonderful success in shallow water, going over bars and riffles where the water was only a foot in depth and making not only time but pulling heavy loads."

If this business stealer wasn't enough to make the blood of Yukon steamboat captains boil, mammoth rafts were also introduced. Construction on the first began in February 1898. It was so large that "there could not be found in West Seattle a strip of level ground long enough" to handle it. When completed, it was 650 feet long and 58 feet wide and towered over the water by some 30 feet. Though the raft had 300,000 board feet, it would float high in the water and thus could be used effectively in shallow-water river areas. This enormous shipping barge would, as well, put a crimp in the conventional river steamer traffic.

But with all the changes, the bottom line remained the same. With the privilege of being the "Gateway to Alaska," Seattle claimed the right to fleece stampeders and wallow in greed. For the *Post-Intelligencer*, it was easy. Whenever a rich man came to town, his name and amount of gold he possessed were printed in the paper. The lists of successful Klondikers ran several columns in the *Seattle Times*. Often, as in the cases below, the information was actually provided by the "purser and captain of the steamer," who, of course, "vouch[ed] for accuracy of the amount."

> *H.M. Kerry, general manager of the Yukon Steamship Company, has $50,000 in nuggets for exhibition purposes; J.B. Rhodes, Portland, Oregon, $60,000 in dust; W. Caldwell, Portland, Oregon, $80,000 in dust (two members of Philadelphia Yukon Syndicate); Edouard Larson, $100,000 in gold dust; Antoin Leoatoad, $100,000 in gold dust, J. Sinclair, Vancouver, $25,000, and some $50,000 divided among the rest of the passengers on the steamer* Fastnet *represents the amount of gold dust brought from the Yukon today.*
>
> *Steamer* Humboldt *arrived in port at 4 o'clock yesterday afternoon, twelve days out from St. Michael. She brought 233 passengers all from up the Yukon, who had among them not to exceed $100,000 in gold dust and nuggets. But the sum was far exceeded by the value of drafts held by returning Klondikers.*
>
> *In the strong boxes* [at St. Michael] *of the big companies is stored over $4,000,000. The yellow metal which has drawn thousands of men to this part of the world has been coming in sacks and boxes by every boat.* [The first three boats down the Yukon brought a total of $5.5 million in gold.]

In a "Special Klondike Edition" on July 17, 1898, the *Post-Intelligencer* listed name after name of individuals and ships and the amount with which they returned. The term "millions" was used time and again, and the steamers bearing the Klondikers to Seattle were called "treasure ships." Adding up the nuggets and dust brought to Seattle during July, the *Post-Intelligencer* estimated the city had seen $7,296,000 pass through its port. (This is about $162 million in 1990 dollars.) Newspapers elsewhere in the nation reprinted these attractive figures, and the unwary flooded to Seattle.

A year later, in July 1899, the paper ran a story quoting William Ogilivie, the "Canadian Government's Yukon Commissioner," who said that the gold coming out of the Klondike topped $100 million in 1898, a pronouncement that did little to stem the tide of Klondikers headed north.

To keep these numbers in perspective, in Seattle in 1898, potatoes were selling at $0.55 a hundred, cheese for $0.10 a pound. Hams were going for $0.09 a pound, bacon for $0.11 and prunes for $0.03. Children's gloves cost $0.50. A dining room table ran $3.75, and parlor sets of three chairs, now called "top quality antiques," went for $24.50 a set. If you wanted to make your own clothes, Scotch gingham went for $0.25 a yard and pure silk at $0.75.

Things were a bit more expensive in Dawson, however. There, a candle cost $1.00. Flour went for $100.00 a sack and oysters for $25.00 a pound. Tobacco cost $7.50 a pound and cigarettes $0.50 a pack. Fresh mutton went for $1.50 a pound, beef at $1.00 and moose up to $1.75. The next summer, oranges and lemons went for $1.50 each and watermelons for $25.00. Dogs sold for up to $400.00, depending on the quality of the animal, and as Adney noted, a copy of Shakespeare's works went for "$50."

But stories of the cornucopia of gold emptying into Seattle coffers said nothing of the thousands of luckless stampeders who returned penniless. By March 1898, the *Review of Reviews* estimated that stampeders spent over $60 million in supplies and tickets, of which Seattle was credited with receiving $16 million. But by the end of the year, the total Klondike output was estimated at only $10 million.

The human toll left by the Klondike gold rush was both staggering and frightening. Of the estimated 100,000 people who actually set out after gold, only 30,000 made it as far as Dawson. About 15,000 of those looked for gold, but only 1 out of every 3 prospectors actually found any. Of these, no more than 300 became wealthy. Those were stiff odds in a luckless game. In modern terms, any ordinary citizen's chances of becoming wealthy from the Klondike rush were about three-tenths of 1 percent. Today, one would have a better chance of making a million dollars on the commodity futures market with $1,000 or picking a winning lottery number.

A single summation of the dreadful human cost of greed and the Klondike rush could not have been better presented than in the July 21, 1898 *Post-Intelligencer*. In reporting the arrival of two groups of Klondikers, the journalist wrote:

> *There are dark shadows to every picture. On Tuesday night, while scores of the fortunate Klondikers who arrived on the* Roanoke *were scattering*

> *gold with a lavish hand in the gilded resorts of the Tenderloin, surfeited with the richest food and flushed with the choicest wines, twelve honest, hard-working men who had been their fellow passengers on the trip down from Unalaska to Seattle, were sleeping in box cars on the waterfront, cold, supperless and without a penny to insure them breakfast when the weary night would end.*

The twelve destitute men were members of the Alaska Mutual Transportation & Mining Association of Chicago. Their savings for years and their gilded hopes and dreams for the future had been shattered by the angry waves of the north Pacific. On the morning of July 4, after a night of terror, the river steamer *Alfred J. Beach*, upon which they were passengers bound for St. Michael in tow of the steamer *Noyo*, was cut loose and settled into the depths of the ocean, a dismantled and water-logged hulk.

Chapter 8

Mass Murder and Robbery or an Accident?

Now that the chronicles of C.H. Lewis and the *Clara Nevada* have been revealed, there remains the speculation of what actually happened on the night of February 5, 1898. Based on documentation I have found, I believe I have pieced together what happened and, to a certain extent, who perpetrated a monstrous crime. What I have been able to develop is a reasonable scenario of events, backed by source material, to support my speculation. In the end, however, what I am presenting is an amalgam of circumstantial evidence.

I now believe that the *Clara Nevada* was deliberately sunk for the purpose of covering up the robbery of $165,000 in gold dust. I believe the thieves set a modest charge of dynamite with a delayed fuse against the inside wall of the boiler room to sink the ship so their robbery would not be discovered. I also believe that the fire was set by the thieves as the best way to keep the passengers and crew distracted until the dynamite exploded. This chapter is dedicated to my speculation and the circumstantial evidence that I believe supports my contentions.

Was the Robbery of the *Clara Nevada* Truly One of the Largest in American History?

Probably. But it is hard to compare recorded robberies simply on the basis of the booty. First, the value of money changes over time. Second, some robbers extract loot that is not convertible, which reduces the overall size of "the take." Third, some robberies are unsuccessful because the loot is lost during the getaway or disappears forever through other circumstances. Finally, there have been robberies so clever that no one even knows they were committed.

In terms of amount of money stolen, America has had its fair share of sizeable hauls. On September 18, 1877, Sam Bass snatched $60,000 in gold from a Union Pacific train near Big Springs, Nebraska. Eleven years later, George Leonidas Leslie dragged $2,747,000 out of the Manhattan Savings Institution. In 1921, Gerald Chapman and two accomplices stopped a mail truck between Wall Street and the New York Post Office and snatched thirty-three sacks of mail containing $1,424,129. In 1893, the Philadelphia Mint was burgled by its chief of weights clerk, Henry S. Cochrane, who got $130,000. And, in 1922, the Denver Mint was robbed of $500,000. In 1950, the infamous Brink's Job netted $1.6 million. There have been other jobs netting the perpetrators into the millions of dollars, probably the largest being the Los Angeles robbery of the Security Pacific National Bank for $10.2 million on October 25, 1978.

Comparing the $165,000 from the *Clara Nevada* to these, however, would be irrelevant. What is important is the distinction between the robbery of the *Clara Nevada* and those listed above. In all but one of the cases above, the perpetrators were captured. In the case of the *Clara Nevada*, up to the date of publication of this book, the thieves had escaped detection.

HOW MUCH GOLD WAS ACTUALLY ON THE *CLARA NEVADA*?

By weight, eleven thousand troy ounces of gold is more than nine hundred pounds of metal. It is logical to assume that the gold that was being brought south by those passengers who lived in Juneau would be in their staterooms. Juneau was only eight hours away, and these passengers may have felt no compelling reason to place their cache in the purser's safe. But those passengers headed for points further south would have been compelled to secure the gold in the safe. Seattle was four or five days away. They could not stay in their staterooms every moment of that journey to safeguard their poke.

Further, the $165,000 figure is for gold alone. At least one newspaper noted that there were "certified checks" onboard as well. A certified check from a Skagway bank in 1898 would probably be accepted in Juneau, but it is doubtful that the same instrument could be cashed in Seattle. Thus I speculate that most of the gold was in the purser's safe. The Juneau passengers were probably carrying a certified check, a much safer means of transporting wealth.

I doubt that the gold was sent as a single shipment. If it was, that would have meant that the entire amount was in the safe. This is unlikely. Most gold was transported on trustworthy ships with regular schedules rather than a tramp steamer like the *Clara Nevada*. Thus, I tend to believe that the gold was in the possession of individuals who had struck it rich in the Klondike.

I also believe it would have taken more than one man to move nine hundred pounds of gold. Lastly, I believe it would have taken a senior crew member to get the gold out of the purser's safe, either the captain or the purser.

Did the Boilers Explode?

I don't believe so. I believe that explosion was caused by a deliberately set dynamite charge. As there was a fireball *and* a blackened hole in the boiler room, it appears obvious that the explosion originated in the boiler room. While there could have been a fireball if the flames had touched off the ship's supply of lamp oil, the conflagration would not have left a single, blackened hole in the boiler room. The ship would have burned to the water line and sunk.

If the boilers exploded, there may very well have been a hole in the boiler room wall, but there would have been no fireball, and it is unlikely that there would have been a "blackened hole" in the side of the ship. As there was both a blackened hole and a fireball, I suspect a charge of dynamite was placed surreptitiously in the boiler room by the thieves.

An Anchorage explosives wholesaler informed me that it would have been very unlikely for the several tons of dynamite to explode without help. Even if the dynamite was burning, it would not explode. An explosion would come from a combination of dynamite with an imbedded blasting cap and a fuse crimped into the blasting cap. After the fuse was lit, the sequence would become lethal.

The explosives dealer also told me that several tons of dynamite was even a large load today, but for a mine, it was probably a year's supply. Today there are strict regulations concerning joint shipment of dynamite and caps, but in those days, it all might have been shipped together. However, even if the blasting caps and fuse and dynamite were all in the same pile and were ignited, it was doubtful that the dynamite would be set off. If they had, all the dynamite would have gone up virtually instantaneously, even though the dynamite was in different boxes. From the size of the hole, the explosives people figured that the charge might have been as small as a stick.

I am inclined to believe that whoever perpetrated the disaster put a modest charge of dynamite, perhaps one stick, against the inside hull wall in the boiler room, lit the fuse and walked away. I doubt that the smell of the burning fuse could have been detected in the stench of the boiler room. If the perpetrator had smelled the fuse, he most likely would have tossed something with a pungent smell into the boiler room fires to cover the odor. He might have even planned ahead and placed a powerful odor producer, like rubber, into the fires before he lit the fuse.

I also suspect that the thieves intended to blow a hole large enough to sink the ship quickly but small enough to allow time for the passengers and crew to abandon ship.

WAS THE BODY FOUND ACTUALLY THAT OF GEORGE FOSTER BECK?

The discovery of the body alleged to be that of George Foster Beck raises perplexing questions. The actual discovery was made in mid-March 1898, six weeks after the *Clara Nevada* had gone down. There were three newspaper stories on the discovery of the body, each of which had information that the others did not. All three put the body on a different island.

According to the story from the *Dyea Trail* of February 20, the body was retrieved by the *Rustler* and had been discovered "off Shelter Island." The corpse had no eyes, and the hair had been "burned off." The body was positively identified by "papers found on it." This article, however, was admittedly incomplete and rushed into print "as the *Trail* goes to press."

The second article ran in the *Post-Intelligencer*. It stated that the *Clara Nevada* had been sunk on Vanderbilt Reef and that the body had been found on "Sullivan Island." "A portion" of Beck's coat and trousers "had been burned," but "the body was in a perfect state of preservation and [Beck's] features could be recognized by all who knew him."[62] Though the reef is unquestionably in error and the island on which the body found most probably in error, it is quite likely that the remaining two facts are correct.

Probably the most accurate account of the discovery of the body was printed the next day in the *Juneau Alaska Miner*. The body was discovered by I. O'Neil on one of the Chilkat Islands. O'Neil did not know that a ship had been wrecked or that a reward had been offered for the corpse. He simply pulled the corpse ashore and covered it with brush until he could report it to the authorities. An examination of the body led to the following inventory:

> *A bank book Washington National Bank with the name G.F. Beck, Treas, P.A.T Co.* [Pacific & Alaska Transportation Company] *on the cover. It contained a report from Washington National Bank for $397.45 in favor of G.F. Beck, Treas., bank checks and a few bills against the* Clara Nevada. *There were some personal items as well, including a bunch of keys on a heart-shaped ring, a small, two-bladed, pearl-handled knife, and two photographs of a lady and child taken together in two different positions…On the third finger of his left hand was a plain gold ring engraved "A.F. 18K."*[63]

Was the body that of George Foster Beck? Probably. But the body's identity is immaterial. It is the condition of the corpse that is vital.

First, the body was burned, indicating that there had indeed been a fire of some kind—physical evidence that clearly supports the burning ship theory.

Second, assuming the body to be Beck's, the fact that it was found at all indicates that the purser must have been on deck when the fire occurred. Obviously, if he had been below decks, the corpse would have gone down with the ship, along with the rest of the passengers and crew.

Third, as I believe that the since the body was found on one of the Chilkat Islands, some survivors may have gotten off the *Clara Nevada* before she entered Lynn Canal.

It would also be in error to assume that a careful search was made for survivors. The rescue mission did not take place the next day. According to the *Sitka Alaskan*, "a day or two" after the wreck, the steamer *Coleman* spotted the wreckage and took the news to Juneau. The *Rustler*, according to the deputy collector in Juneau, left to search for survivors on February 9. That means there were at least three days between the sinking of the *Clara Nevada* and the beginning of the search. By then, any survivors that had made it ashore would be on foot, trying to make it to civilization. Skagway was only thirty miles up the coast, and there were surely smaller vessels plying the waters looking for survivors as soon as the storm abated. Survivors who ended up on the mainland side of the channel could have easily walked to the communities of Comet or Jualin, no more than a few miles from where the *Clara Nevada* debris was found. As Comet had left its light blazing the night the *Clara Nevada* went down to give any survivors a beacon toward which to steer, anyone who made it ashore would have known in which direction to walk. They could not have missed the Comet dock if they were walking along the edge of the inlet.

Most likely, the *Rustler* was looking for people standing on shore waving to be rescued. According to the search record, the *Rustler* steamed along the coastline "close enough to distinguish objects with a glass but could see no one or discover any signals of distress." Further, according to the *Alaska Searchlight*'s February 12 edition, "not a sign of life was to be seen, and heavy snows have fallen since the wreck which would have covered up any of the dead that might have been cast ashore by the angry waves."

But the supposedly careful search by the *Rustler* raises other questions. If the search for survivors had been competent enough to spot the tracks of a dog on shore, why wasn't it good enough to find the lifeboat? That was well within the area searched by the *Rustler* and probably right on the shoreline.

And when the *Rustler* visited the community of Comet, why didn't someone tell them that there had been survivors? After all, George Hemmings was there for the 1900 census. It would seem logical that he was there shortly after the wreck.

If the Body Was Indeed That of George Foster Beck, What Does the Body Indicate?

I am inclined to believe one of two scenarios in regard to George Foster Beck. On one hand, he could have had some part in the conspiracy to steal the gold. Why, after all, would he, the purser, have been on deck in a ninety-mile-an-hour snowstorm? Further, he had access to the gold and the opportunity to spirit the precious metal out of the ship's safe.

On the other hand, Beck may have become suspicious of some members of the crew. They were, after all, an unsavory lot. Beck might have come on deck, discovered the robbery in progress and been knocked unconscious, perhaps even killed before he could spread the alarm. He would have lain on the deck, dead or alive, until the fire reached him. That he was burned and not blown apart would seem to indicate that his fate was sealed before the dynamite sank the ship.

Is There Any Evidence to Support the Contention That George Foster Beck Was Involved in the Robbery?

To assume that George Foster Beck had a hand in the disaster is speculative. After all, he was a young man with a new family who was the heir to a large estate in Oregon. Why would such a person stoop to robbery and murder? Is there any evidence to suggest that he had such a proclivity?

Perhaps.

According to the published letter of Latty Boyce, the carpenter's assistant who made the trip north to Skagway, George Foster Beck was a "willing party" to a petty swindle involving lighterage of passenger's cargo. This amounted to "$2 per load" for passengers' baggage and "$5 per ton" for freight. There is no way to translate this into any kind of a meaningful figure. If all three hundred tons were involved, the total would have been in the range of $1,500 for Beck and Rogers, the freight clerk.

There is no corroborating evidence to this statement. It appears not to be true because Beck was found with very little money in his pocket. If he was robbing the ship and planning on leaving the vessel, why wasn't he carrying the money with him? Either the swindle was very small, or Beck left the money in his stateroom and it went down with the ship.

Another passenger, J.W. Lansing, M.D., wrote a letter to the *Seattle Times* in which he stated that Beck had confirmed owning one-third of the Pacific & Alaska Transportation Company. Further, Beck and the freight clerk appeared to be the "power behind the throne" and exercised their power frequently. If the captain didn't recognize them as such, the crew certainly did. As far as Beck's greed was concerned, "[Beck and Rogers] inspected our filthy, cold, dark compartments, but claimed it was good enough fare for second class passengers…Almost everything about the steamer gave evidence of cheap botchery." By contract, another passenger, Charles Jones, found "the captain, purser and freight clerk attended strictly to business and were gentlemen."

With regard to the poor condition of the ship, according to Boyce, "Mr. Beck and Mr. Rogers [the freight clerk] wined and dined the Government Inspector at intervals all through the night." Boyce's implication that Beck was intent on diverting the inspector's attention from the state of the vessel is realistic, considering the poor condition of the *Clara Nevada*.

In all fairness, it should be pointed out that when the *Post-Intelligencer* ran an article on the crew who had perished in the disaster, George Foster Beck was listed as "one of the best-known young men in Portland having lived there the greater part of his life...and the son-in-law of Mrs. R.L. Hawthorne, probably the wealthiest woman in Portland."

If indeed George Foster Beck had been involved, it would be logical that his most likely accomplice would be George Rogers, the freight clerk. While I could find no evidence that Rogers was alive and in Alaska in 1900, the *Post-Intelligencer* noted that "it is possible the freight clerk, George Rogers, may have remained in Skagway and it is also learned that several members of the crew intended to desert in Alaska." Rogers had been "instructed to remain at Skagway to act as an agent for the company."[64] Whether or not Rogers was on board when the vessel went down is open to speculation.

Interestingly, I did find a reference to a George Rogers in the records of the United States Assay Office in Seattle. A man by that name brought in 79.83 ounces of gold, about $30,000 today. The origin of the gold was written in an abbreviation that was not decipherable, but it did not come from Nome, as that abbreviation was clearly readable.[65]

It is also important to note that a ring of keys was found on Beck's body. Those were most likely the keys to the various padlocks and doors on the *Clara Nevada*. That Beck had the keys proves nothing one way or the other. Had he been involved in the robbery, he would have maintained possession of the keys. Had he not been involved, it would be grounds to assume that Lewis had been the officer who was the perpetrator.

WHAT ABOUT THE LIFEBOATS?

Facts about the number and size of the lifeboats are of interest. According to Boyce, there had been only four seaworthy lifeboats. "Two of them were the old ones on the *Hassler* and two were new iron boats." A fifth boat, a skiff, was on the stern, but it "was just simply an old rotten skiff which would not have floated to be rowed across Elliott Bay." In fact, this skiff was in such bad shape that passengers were told that it had been "shipped as freight."[66] As far as the printed record is concerned, only one of those boats was on the wreckage of the *Clara Nevada*.

The *Post-Intelligencer* reported that two of the lifeboats were "metallic with water-tight compartments" and had not been found. This seems to tally with Boyce's statement that two of the boats were iron.[67] Lewis, on the day before the *Clara Nevada* left Yesler's Dock in Seattle, complained that the inspectors had forced him to get "metallic lifeboats."[68] These must have been the ones in question.

Yet, according to the *Dyea Trail* article, the lifeboat that was found not far from Eldred Rock was "neither a lifeboat nor a sealing boat." It did not say it was "metallic" or "iron." Thus the description of the boat that was discovered does not match any of the boats described by Boyce. It is unlikely that Boyce, a maritime carpenter by trade, would have mixed up his boats. Boyce had been working on board since "January 13" and had thus spent three weeks on board, certainly enough time to have discovered the additional lifeboat.

Then there is the name "J.H. Burns, S.F. builder" found on the discovered lifeboat. This may be significant in that the lifeboat apparently came from San Francisco.[69] While it was not unusual for steamers at that time to have lifeboats from anywhere along the West Coast, it seems odd that such an unusual lifeboat would have been on a steamer, and the ship's carpenter would know nothing about it.

It is my speculation that the lifeboat discovered on the beach near Point Sherman was brought on board in Skagway, after Boyce left the ship. The fact that it was unquestionably seaworthy leads me to suspect that someone on board was interested in a sturdy and dependable escape route.

As to the number of lifeboats, there were six on board. One was found on the wreck, one was unserviceable and one was found on shore. That leaves three missing, which could have carried survivors. If it is assumed that the thieves took one, two are still unaccounted for. According to the man who headed up the initial search, Inspector T.A. Marquam, deputy collector of the United States Customs in Alaska, he found "parts of broken boats" on the shore.[70] If the term "boats" implies two, then all lifeboats are accounted for if the unserviceable boat is assumed to have gone down with the ship.

I checked local newspaper files for months following the wreck to see if there was any mention of lifeboats being found. No article on the lifeboats appeared in the local papers that still existed. However, there were huge gaps in the newspaper runs, many of them immediately after the wreck. There was no mention of any such discovery in the Seattle papers either. If a boat did survive the wreck, I suspect that it was one of the "metallic" ones. It would probably have survived the wreck and certainly the sea. But I am also inclined to believe that the storm would have blown it quite a ways south. Considering the thousands of miles of shoreline and the low density of population in Southeast Alaska, it is also possible that the lifeboat may have been blown ashore and camouflaged with debris.

WAS THE *CLARA NEVADA* ROBBED?

I believe that the circumstantial evidence supports the contention that the *Clara Nevada* was robbed. I am inclined to believe that, in Skagway, it came to the attention of a crew member or officer that there was going to be $165,000 in gold on the ship. The three most logical suspects are the freight clerk, purser and captain: George Rogers, George Foster Beck and C.H. Lewis. All three had the greatest *opportunity* to carry off this caper.

- George Rogers, the freight clerk, would have been aware of all passengers coming on board and what freight they were loading. An extra lifeboat could not have been brought on board without at least his knowledge. Further, he was apparently ordered to stay in Skagway as an agent of the company. Whether he was legitimately assigned to stay in Skagway or his stay was arranged by the rest of the thieves, he would have been available to handle whatever transportation was needed to pick up the thieves once they had done their deed. Rogers also had the connections to get control of the purser's key—if the purser was not involved in the theft.
- George Foster Beck, the purser, had access to the purser's safe, where the bulk of the gold was presumably stored. He and Rogers were friendly and, if the accounts of some of the passengers correct, were already in cahoots on some minor chicanery.
- Captain C.H. Lewis had access to the purser's safe as well and control of the entire ship. It is doubtful that an extra lifeboat could be brought on board and placed on deck without his knowledge.

Robbing the vessel must have seemed a relatively simple operation. The perpetrators would simply open the safe and move the gold to the new lifeboat on the deck of the ship. Just before they cast themselves adrift, they would start a fire onboard the *Clara Nevada*. In the confusion, they would not be missed. Passengers and crew would abandon ship. Events would be moving so fast that no one, passenger or crew, would be able to keep track of who was where as the boats were lowered. Then, when the survivors gathered in Skagway a few days later, the thieves would simply say that they had abandoned ship with everyone else.

As far as the *Clara Nevada* was concerned, it would fill and sink in deep water. No one would even know the gold was missing. Even if the ship could be raised, it would take years. By then the perpetrators would have died of old age, if they could even be linked to the crime. More likely, it would be assumed that a salvage operator had surreptitiously removed the treasure.

Further, while the lifeboats on the *Clara Nevada* were hardly seaworthy, they would have been sturdy enough to get the passengers and crew ashore. At its greatest distance, the Lynn Canal measures only four or five miles wide. In the Chilkat Inlet, where I believe the thieves left the *Clara Nevada* with the gold, the channel narrows to a mile. It grows to two miles in width until the Chilkat Peninsula disappears into the sea on the west.

But the perpetrators on board needed some assurance of survival. They wanted a boat that they knew was sturdy enough to take the nine hundred pounds of gold, plus their combined body weight, and make it to shore. Thus, a sturdy craft that was "neither a lifeboat nor a sealing boat" was probably taken onboard with the assistance of the officer.

Just as the *Clara Nevada* left port, one of the conspirators must have slipped into the boiler room to set the charge. He probably lit the fuse as well. Then, as soon as the *Clara Nevada* was out of port, the perpetrators began moving the gold into the new getaway lifeboat. But there they got a surprise. At least three stowaways were hiding in the craft: William Hemmings, George Kasey and George Logsdon. At least the stowaways had had the good sense to choose the best lifeboat in which to hide!

With the fuse lit and time at a premium, the conspirators could hardly have spent precious minutes arguing with stowaways. Since they probably didn't want the stowaways to know what they were doing, they shifted their attention to another lifeboat.

Another reasonable possibility was that the stowaways could have been armed.

The trip and transfer of the gold to the other lifeboat must have been difficult. First, the ship's timbers were probably coated with ice from the storm. Second, waves were undoubtedly washing the deck. Third, the ship was pitching violently. Fourth, moving nine hundred pounds of metal in these conditions must have been, at the very least, a Herculean labor.

At this moment, there may very well have been another complication: George Foster Beck. Either he was getting cold feet and did not want to conclude the robbery, or he had come on deck to investigate suspicious circumstances. In either case, he was undoubtedly knocked to the deck. There he lay until his corpse or unconscious body was rolled overboard.

Odd as it may seem, there may have been a witness to the machinations of the thieves on deck. In spite of the fact that the *Clara Nevada* was running with a wind estimated to be as high as ninety miles an hour with towering waves and a snow storm, figures were seen on deck by a passing ship. When the *Wolcott* passed the *Clara Nevada*—the former going north against the wind and the latter south—Captain Donaldson of the *Excelsior*, undoubtedly onboard as a passenger, reported that the decks of the doomed ship were almost deserted. Several figures were seen, and they waved to us. That must have been just an hour or so before the occurrence of the explosion, if such really took place. Does it not seem improbable that no one has escaped to tell the tale? Here we have a qualified eye witness who saw figures on the deck of the *Clara Nevada* in miserable weather an hour before the explosion.

Just before the thieves escaped, they set fire to the ship to keep the crew and passengers occupied. This contention seems to be supported by the fact that an early investigator to the wreck, Sanderson Reed, stated that he was certain that the fire had started in "different parts of the vessel."[71]

Had the fire been confined to one area, I could have been convinced that it could have been an accident. But, as the fire started in many locations, I believe those fires were set on purpose. Those fires must have been ferocious, too. According to the eyewitness on shore, "sudden flames were seen to burst from every part of the ship." The ship burned for "20 minutes when there occurred a terrific explosion which [could be] plainly heard above the howling of the win [*sic*]."

Considering the storm and the monstrous waves, the corpse of George Foster Beck may have been rolling back and forth across, or into and out of, the flames quite a number of times, which charred "a portion or his coat and trousers,"[72] but did not incinerate the remains. From Flat Bay to the Chilkat Islands, about six or seven miles or twenty or thirty minutes if the ship were going fifteen knots, there was more than enough time to scorch the cadaver before it was washed overboard by the towering swells that were rocking the ship and sweeping the deck.

Had Beck's body been discovered on Sullivan Island, the corpse would have had an additional ten minutes of burning or may have been swept off the deck when the ship exploded. I doubt that Sullivan Island was the repository of George Foster Beck's body, however, for the wind was blasting out the north. The body would have ended up with the rest of the *Clara Nevada* debris, to the south. Sullivan Island is due west of Eldred Rock.

I also discount Shelter Island as a possibility where Beck's body might have been discovered. Shelter Island is twenty miles to the south of Eldred Rock. For a body to have made it as far south as Shelter Island, the corpse would have had to have missed Point Sherman seven miles away, where most of the debris was found, then stay midchannel for the next twenty miles. More than likely, if any debris had made it around Point Sherman, it probably would have been beached long before it made a few miles down the channel.

I suspect that the perpetrators abandoned the burning ship with the gold between Battery Point and Flat Bay. The ship would have been close to the Chilkat Peninsula and somewhat protected from the blasting north wind. Just as likely, the rendezvous point was either Flat Bay or Seduction Point. Both are somewhat protected from the wind and close enough to Skagway to return easily once the storm passed.

As far as the burning *Clara Nevada* was concerned, I'm inclined to believe that whoever was steering the ship stayed at the wheel until the very end. When the vessel passed the first of the Chilkat Islands, Talasani, the protection of the Chilkat Peninsula would have disappeared and the flaming vessel would have been running before the full force of the ninety-mile-an-hour storm. If the fire was as ferocious as described, it is hard to believe that anyone was left alive below decks by the time the ship rounded Talasani Island.

As the ship passed the last of the Chilkat Islands, whoever was steering the ship must have known that he was doomed. A fire was raging out of control below deck, and the storm threatened to sweep the ship into the shoals. Yet, by looking at a map, it is apparent that the ship ran almost due south until it hit the reef a few hundred yards north of Eldred Rock. As there was no lighthouse there in 1898, a pilot could not have seen the rock in the darkness.

Since the ship did hit, explode and sink at Eldred Rock, I am inclined to believe that the ship was beached on purpose. As it would have been the first watch, I assume that pilot Ed Kelly would have been steering the ship. Kelly knew of the island, as he had inched past it on his way up this channel just the day before. Kelly probably felt that his only chance of survival was grounding the ship. But before the ship hit the beach at Eldred Rock, the explosion occurred and a blackened hole appeared in the boiler room's hull wall. Then the ship went down quickly.

I disagree with Sanderson Reed and the Steamboat Inspection Service as to the cause of the explosion. They both felt that the shock of the ship hitting the rock caused lighting oil to spill, thus precipitating the flash of fire. C.F. Stagger also confirmed that the *Clara Nevada* had "slipped partly off the rock and at high tide is about twenty-four feet under water." But this statement might be misleading, as Skagway can have thirty-foot tides.

I could believe that the *Clara Nevada* may have hit a rock on Eldred Rock's shoals, but she must have been in a trough when the collision occurred. The ship may have hung up momentarily, but it did not remain stationary long.

Further, considering that the channel was experiencing heavy waves that night, it is unlikely that anyone would have seen the fireball had the ship been at the bottom of a trough. The ship more likely was on the crest of a wave when the dynamite exploded. Then, according to estimates, it sank in "three minutes," at least according to the *Seattle Times* of February 16.

At least one of the witnesses claimed that the *Clara Nevada* was stationary before she blew. This could be an accurate statement considering the angle from which the eye witnesses were watching. From Comet, the ship would have been moving directly toward them or at a very slight angle, whether she was being steered or driven by the wind. From that angle, a witness would have no depth perception. As an illustration, consider a person standing in the end zone and watching a football player run directly toward him. From the position of the viewer, it is not possible to tell if the player is actually moving forward or simply running in place.

Additionally, if the ship did indeed sink in three minutes, the only people who could have survived with that kind of escape time were those who were on deck when the explosion occurred. The most likely people to survive would have been the stowaways. They were already in a lifeboat. It would also be my guess that this lifeboat was only secured to the deck by a rope. There would have been no housing braces for the lifeboat, as it had been brought onboard earlier that day. This must have made it easy for the stowaways to cut it free.

Depending on when they started cutting the boat free, they could have found themselves afloat immediately. There was probably a very limited suction, as the ship was only sinking in twenty-four feet of water. The miracle was that the stowaways were able to get the boat launched at all.

The only other person who could have gotten into this lifeboat would have been the pilot, Ed Kelly. He would have been standing in the pilothouse, about one-third of the way back from the bow. Assuming that the lifeboat would have been on the stern of the vessel, even if Kelly had known that the stowaways were launching the boat, he would have had to run one hundred feet down an icy deck, brace himself against the turbulent waters that were washing the deck and hang onto the railing as the ship crested each wave. That he never made it is understandable.

Then there are the lifeboats. All appear to be accounted for. As far as survivors among those who were below deck, I find it hard to believe that anyone below deck could have survived the fire and the explosion and still made it onto the deck, run as far as thirty yards on a slippery deck, gotten into a lifeboat and set it free into fifteen-foot waves within the three minutes it took the ship to sink. I do not believe that anyone survived the burning and explosion other than the three stowaways and the thieves.

Thus ends the saga of the *Clara Nevada*.

What About C.H. Lewis?

Now—with George Foster Beck dead and George Rogers untraceable—what about C.H. Lewis? If he was innocent of any part of the sabotage, why did he not make an appearance in Skagway and tell his side of the story?

A better question is how he made it back to Seattle without being recognized. After all, he was a well-known sea captain. so it would seem unlikely that he could have just walked aboard another vessel unseen or unrecognized. He was, after all, headed south. The gold rush fleet was loaded going north. Any passenger headed south, particularly in February 1898, would have come under the scrutiny of the crew, if not the captain of the vessel. He may very well have assumed a disguise.

Further, it is just as unbelievable that he could buy a new steamboat without arousing some kind of suspicion and then end up in court suing the owners of the *Noyo* without someone making the connection between him and the *Clara Nevada*. After all, Lewis was listed in all the newspapers as having gone down with his ship, so his sudden appearance in Seattle must have raised someone's eyebrows. But even the McGuire brothers didn't know that he was alive. The wreck report for the *Clara Nevada* that was signed on April 22, 1898, specifically listed Lewis as having gone down with the ship.[73]

Another good question is why Lewis buried his ownership of the *William H. Evans*. It took me almost two years to find some paperwork on the vessel. It was listed as having been registered in Baltimore, but I could find no further information. I later discovered it was also registered in Port Townsend, Washington. The corporation that bought the ship was located in West Virginia.

I do not believe that Lewis was trying to hide the ownership of his company by incorporating in West Virginia. More than one Alaskan mining company was registered in West Virginia, as was the Alaska Steamship Company. The laws of West Virginia must have been particularly favorable to corporations at that time. Thus there was an entrepreneurial reason for the choice of West Virginia for incorporation.

There is even a reasonable chance that Lewis was not even on board the *Clara Nevada* the night it went down. J.W. Lansing's letter to the *Seattle Times* states that there was a rumor that Lewis was soon to leave the *Clara Nevada*. No other source confirms this rumor. However, I feel it would have been unlikely for Lewis to abandon his ship in Skagway. If he had planned to skipper his own boat, he would have wanted to leave the *Clara Nevada* in Seattle, not Skagway. Additionally, according to the *Victoria Daily Colonist* on

February 23, Lewis had "tried his best to obtain a release from command at Juneau," but there is no indication he succeeded.

Further, if Lewis had not been on the *Clara Nevada*, I believe that fact would have been mentioned in the newspapers. He was, after all, a seasoned mariner and would have been known by most of the sea captains on that route, if not by many members of the crew of the various steamers as well. If he had not been on the *Clara Nevada*, at least one of the ships coming south would have brought that fact to the attention of the newspapers or maritime authorities; after all, the hearings into the *Clara Nevada* lasted well into August 1898, and the Seattle newspapers were using the case to bludgeon each other. If Lewis were known to be alive, how could that fact *not* have been brought out?

I am also inclined to believe that the Lewis Klondike Exploration Company had already put down earnest money to buy the steamship that was to be named the *William H. Evans*, and that Lewis was just biding his time, waiting for the ship's construction to be completed. It was probably Lewis's intention to return to Seattle on the *Clara Nevada* and take command of his new ship immediately.

As to the money to buy the *William H. Evans*, where did Lewis, the survivor of a shipwreck, find the cash needed to buy a brand-new steamer? A new ship was expensive. In February 1898, a ship that was 125 feet long, 30 feet wide and 6 feet deep cost $230,000. While Lewis may have had access to almost $400,000 with his corporation, that money was on the East Coast. The *Evans* was purchased on the West Coast in February 1898.

Further, even though he had access to the entire $400,000, Lewis apparently chose not to use this money to pay for the *Evans*. In fact, I find it doubtful that Lewis left the West Coast after the disaster. Instead of tapping his company's reserves, he found new investors, many of them very small. Thus, I suspect that one of two things happened. First, and least likely, the Lewis Klondike Exploration Company had some of its incorporators abandon the project. This would have left Lewis short on cash. If this falling out had occurred *before* the *Clara Nevada* left Seattle, Lewis would have had a strong motive for murder and robbery.

Secondly, and more likely, Lewis decided to raise the money himself on the West Coast. This would have saved him the investment of his own capital, as well as spared him the hurried trip east to bring some cash west. I believe that he raised the money on the West Coast. Since the firm was renamed the Evans Klondike Exploration Company, I therefore believe that Evans was the majority stockholder. If the ship had cost in the neighborhood of

$250,000 and Evans owned half the venture, the reason that the company was renamed, that only left $125,000 to be raised.

It would be my guess that construction on the *William H. Evans* was initiated with a down payment from Lewis in late 1897 or early 1898, 10 percent being a reasonable amount. This would mean that the twenty-five small investors would be paying the balance of $100,000. Individually, they would have had to pay $4,000 apiece, $100,000 in today's dollars. This would not have been an unreasonable amount to expect from a "small" investor.

Lewis, in essence, probably did not drop a penny of his own money into the *Evans*. I further surmise that when he was tabulated in the 1900 census in June in Baltimore, he was withdrawing money from his West Virginia company to pay the $13,000 to salvage the *Evans*. Thus, when the books were closed, I suspect that Lewis himself only paid about $38,000 for the entire *Evans* imbroglio. Even if Lewis had absorbed all of that loss, he would have had $262,000 left of his original investment. He was still a rich man; everyone else on the *Evans* was left destitute.

While I am reasonably certain that Lewis was involved with the robbery and murder, I am just as certain that he did not profit from the crime. There is no solid financial evidence that he did. In fact, I feel that the acquisition of the *William H. Evans* was probably done with wheeling and dealing, not with stolen gold.

After examining the evidence, I felt that there must have been at least one more person involved. While George Rogers might have handled the escape, and George Foster Beck or C.H. Lewis might have had access to the purser's safe, there was still a tag end to be resolved. While these men may have been a bit bent, they were not mass murderers. For this very reason, I believe that they were involved. The setting of a modest charge of dynamite indicates that there was concern for the passengers and crew. A more unsavory individual would have set off the entire cargo hold of the explosive if such were on board.

I think there was possibly a fourth person involved, and I believe that person to be the fireman, Paddy MacDonald. I believe that the Paddy MacDonald of the *Clara Nevada*'s crew was also the same "all round crook" that the Royal Canadian Mounted Police included in their list of "the worst criminals known on this continent," and he could very well be the old-timer who was mentioned in the *Teller News*. I believe this for two reasons.

First, I find it hard to accept that there could be two Paddy MacDonalds listed in historical documents at about the same time in the same general area. Paddy MacDonald is, after all, a very unusual name. "Paddy" is an Irish name of endearment, yet MacDonald is Scotch. There were probably not a lot of Scots walking around Alaska with a first name of "Paddy." Second, I could find no evidence of any other Paddy MacDonald in the historical record in the same time period.

If these two Paddy MacDonalds were the same person, he may have had no qualms about blowing up an entire ship to escape with $165,000. Being a fireman, MacDonald would have had access to the boiler room and could have easily set the charge. I also suspect that he was the one who set fire to the ship, an action I find hard to believe of George Foster Beck or C.H. Lewis. Reflecting upon the barbarity of the crime, this may have been the cause of the dispute with Beck on the heaving deck of the doomed ship.

I believe that Paddy MacDonald was in a perfect position to set the charge. He had access to the boiler room, and his presence there would not have aroused suspicion. The presence of the captain or the purser in the boiler room, on the other hand, would have raised eyebrows on the part of the other boiler room crew.

Traveling to Nome after Dawson was probably a smart move on MacDonald's part. Dawson was clearing out, and Nome was booming. Like Dawson, Nome had all of the advantages of a gold rush town: lots of sheep to be fleeced and a huge flock into which he could blend. Nome was about two blocks wide by five miles long and, in 1900, had a population estimated at twenty-three thousand. With crowds like this, it would have been very easy to remain anonymous.

What Happened to the $165,000 in Gold?

I could find no news article that indicated the safe had ever been recovered. Thus I believe that the gold never made it to shore. I checked with corporate records in Oregon and Washington and found nothing under any name that was familiar. The records of C.H. Lewis do not indicate a partnership with any former member of that crew.[74]

If I had to say where I believed the gold could be found, I would say it was fairly close to shore somewhere between Flat Bay and Seduction Point. I am inclined to believe that the gold would be on the western side of the channel, since the Chilkat Peninsula would have given the lifeboat protection from the wind. However, if the gold were to be found on the eastern shore of the channel, I would not be surprised. The lifeboat may have run with the wind, and depending on the angle of the blast, the lifeboat could have been swept eastward.

Of course, where the lifeboat actually swamped is speculative. The thieves quite likely loaded the gold into the lifeboat and made for shore. But I doubt that they made it. They were in an old lifeboat in heavy seas with nine hundred pounds of gold weighing them down. Most likely the weight splintered the boat close enough to shore for the men to swim to safety.

Even if they were in one of the metallic lifeboats, the trip would have been no less treacherous. Those lifeboats were fairly long, and with only two men aboard, it would have been hard to row, much less navigate through fifteen-foot waves. Further, it is quite reasonable to assume that a metallic boat would have been just as sluggish in the pounding surf near the shore as a wood boat.

Depending on where the gold went down, it could be in water as shallow as thirty feet or as deep as three hundred. But, I suspect, the gold is in fairly shallow water. Whoever was in that lifeboat had to be able to make it to shore through twelve- to fifteen-foot waves and ninety-mile-an-hour wind. Had the lifeboat swamped too far from shore, the thieves would never have made shore.

While the newspaper says that the gold was in dust form, I am fairly certain that it was boxed. If a man was coming onboard with $30,000, for instance, he would not have walked on board with 125 pounds of gold in a canvas bag. He would have brought his bounty aboard in several wooden crates. That would have made it easier to carry. Even more likely, the gold would have been in doré bars, crude bullion.

With regard to the physical size of the boxes, the larger ones were about two feet long, a foot high and possibly six to nine inches wide and bound with steel hoops. I am making this estimate based on an examination of such boxes in a front-page photograph, perhaps the first half-tone ever published in the *Post-Intelligencer*, on July 19, 1899.[75] Smaller amounts would have been placed in smaller boxes.

But the important consideration is that the gold dust would not have been loaded loose. It would have been loaded in boxes that would have made it easier to handle. When the lifeboat sank, the gold in the boxes sank as well. I'm also inclined to believe that when the gold went down, all of it hit the bottom in the same general area.

From an examination of the nautical chart, I would say that the area where the lifeboat sank could best be described as a ledge above the steep slope of a submarine mountain range. The Chilkat Inlet is, after all, an underwater fjord. But it does have a lip. There appears to be a shelf of about 100 yards in width down to about 15 fathoms (90 feet), running along the side of the Chilkat Peninsula for the 4 miles between Flat Bay and Seduction Point, but that shelf is very narrow, possibly no more than 100 yards wide. Then there is an abrupt drop-off into water that is 80 to 100 fathoms (480 to 600 feet) deep. On the east side of the channel, the shelf probably runs about 8 miles and is also no wider than 100 yards.

However, I think the gold is in the shallow water. The lifeboat had to have been within one hundred yards of shore for the survivors to have made it to dry land. I also believe that the waves were probably more vicious in the shallower water and that must have been the deciding factor.

While the boxes may have long since rotted away, the gold is most likely still there, all eleven thousand troy ounces of it, waiting for some lucky skin diver to stumble upon it.

Whoever finds that gold, if it is actually there, will be the one to write the last chapter on the saga of Alaska's ghost ship, the *Clara Nevada.*

Appendix

Partial List of Passengers of the *Clara Nevada* Drawn from Alaskan, Californian and Seattle Newspapers

Banks, Robert Bruce[76]
Bonicke, F. (Juneau)—also spelled Benicke
Brenneker, A. (Portland)
Hemming, William—probably a stowaway and a survivor
Hill, Thomas R.
Holgate, E.T. (Portland)
Hunt, Harry (Montana)
Kasey, George—probably a stowaway and a survivor
Lee, Dr., and wife
Logsdon, George—probably a stowaway and a survivor
Londer, M.—could also be Lander or Linder
Malloy, William—corpse being transported south[77]
Mead, Dr. C.A.—reported lost on the *Clara Nevada* in the *Olympia Washington Standard*, February 25, 1898, but "returned home on the 23rd;" therefore, he probably was not a passenger or a survivor
Noois, Colonel A.
Novis, A.
Noyes, Al (Juneau)—this man and the two above could be the same man
Ross, A.P., wife and child
Saportas, E.W.
Selang, A.J. (New York)

Smith, ?
Turner, Frank (Colorado Springs)
Whitney, Frank (Seattle)
Wilkins, Jesse (Portland)
Wroe, George—could be George Roe, a crew member
Eskimo baby[78]

CREW OF THE *CLARA NEVADA*[79]

Beck, George Foster, purser—only body recovered (Portland)
Benton, H.M., night watchman
Bowen, Harry, second officer (San Francisco)
Bowman, Frank, mess boy (Connecticut)
Butler, John L., cabin boy and waiter
Carey, Billy, fireman
Cunningham, Dan
Dehm, Earnest "Doughnuts," porter (Seattle)
Emery, Frank, steward's assistant[80]
Finnigan, Arthur, fireman
Gibson, ?, second officer
Hill, George G., steam fitter
Hunt, Charles, assistant steward
Hurley, Bat, quartermaster
Jackson, W.J., steerage steward
Jacobs, W.A., carpenter (Portland)
Kelly, Ed, pilot (Port Townsend)
Kelly, ?, waiter
Lewis, C.H., captain—survivor
MacDonald, Paddy, fireman—possibly a survivor
Moser, ?, assistant engineer (Seattle)
O'Brien, Edward, quartermaster
O'Donnell, Frank, steward—reported to be "on one drunk the whole trip through"[81]
Perkins, C.E., cabin boy and waiter
Perkins, Walter
Reed, David, engineer (San Francisco)
Roe, George, cabin boy (Tacoma)
Rogers, George, freight clerk—may not have been on boat[82]
Williams, Thomas, first assistant engineer (San Francisco)

Notes

1. The Maritime Rush to Skagway and Dyea

1. *Post-Intelligencer*, "Balloon is Completed," February 17, 1898; *Post-Intelligencer*, "Reindeer to Carry Mail," March 30, 1898; *Post-Intelligencer*, "Making Ready for Alaska," March 15, 1898; *Post-Intelligencer*, "To the Klondike by Canoe," March 18, 1898; *Post-Intelligencer*, "Camels for Icy Alaska," February 24, 1898; Terrence Cole, *Wheels on Ice* (Alaska Northwest Publishing Company, 1985); *Post-Intelligencer*, "Making Ready for Alaska," March 15, 1898; *Post-Intelligencer*, "Fatal to Reindeer," March 28, 1898; *Post-Intelligencer*, "The Famous Tin Boat and her Scandinavian Owners," November 5, 1898; *Post-Intelligencer*, "Want to Fly to Klondike, January 22, 1898; *Post-Intelligencer*, "Reindeer for the Camps," August 24, 1897; *Post-Intelligencer*, "The Aerial Voyager" (with freehand drawings), August 18, 1897.
2. McCurdy, *The H.W. McCurdy*, 18.
3. *Post-Intelligencer*, "Say They Were Kidnapped," September 20, 1899; *Post-Intelligencer*, "Prayed on the Icy Deck," July 5, 1899; *Post-Intelligencer*, "Terrible Tale of the Sea," February 21, 1899.
4. *Post-Intelligencer*, "From Inspector Dument," February 16, 1898.

2. The Wreck of the *Clara Nevada*

5. I could not find a specific location named Seymour Narrows on any current map. Because of similar geographic phrases, I assume it to be in

British Columbia, probably in the Queen Charlotte Strait area, between the northern end of Vancouver Island and the mainland.

6. *Post-Intelligencer*, "Steamer *Eugene* Is Abandoned," September 7, 1898.
7. *Post-Intelligencer*, "Steamer Libeled," February 9, 1898.
8. *Pony Express*, "*E Clampus Vitus* Barnstorms Old San Francisco," March 1940; *Pony Express*, "Once Historic Landmark of the Comstock," June 1940; *Pony Express*, "The Great Emma Nevada," April 1941; *Pony Express*, "Hell-Fire Bill from Gold Hill," July 1945. William "Billy" Metson's life was so colorful that it was used as a model for the character Billy Wheaton in Rex E. Beach's immortal novel *The Spoilers*.
9. *Seattle Times*, "*Clara Nevada* Rumor," February 15, 1898.
10. Wreck record in the Alaska Packers Association Records, Z.J. Loussac Library, Anchorage, Alaska. For the historian interested in details, the record was signed by H.P. McGuire, president of the Pacific & Alaska Transportation Company. The vessel's number was 127222. W.W. McGuire, the other half of the McGuire brothers, died in 1937. An obituary for W.W. McGuire can be found in the *Oregonian* on December 19, 1937. Of passing interest, the *Clara Nevada* was reported to have "yellow trimmings," according to the *San Francisco Call* on February 16, 1898.
11. *Seattle Times*, "Another Phase of It," February 23, 1898.
12. *Sitka Alaskan*, "Loss of *Clara Nevada*," February 19, 1898. To be consistent with current maps, all references to the community of Seward City will be listed as Comet.
13. Deputy Collector C.S. Hannum to Collector of Customs, Sitka, February 18, 1898, letter, Alaska State Library, Juneau. Of passing interest, the debris was still on the shores of the canal the next year because a portion of the ship's rail was turned into a walking cane, (*Seattle Daily Times*, "Moran Fleet," April 4, 1899).
14. *San Francisco Call*, "Fruitless Search for Survivors," February 16, 1898.
15. *Post-Intelligencer*, "Saw *Nevada* at Juneau," February 15, 1898.
16. *Seattle Times*, "Divers Find the Hole," February 25, 1898.
17. Newspaper accounts of the actual number of passengers varies. The official wreck report, filed April 25, 1898, lists passengers as "from 30 to 40" and crew of "about 42." The wreck report can be found in the Alaska Packers Association Records, microfiche, Z.J. Loussac Library, Anchorage, Alaska. The *San Francisco Call* noted in "Vain Quest for Bodies of the Lost" on March 5, 1898, that "it is the rule that bodies of people drowned in Alaska seas are never recovered. Once drowned they never rise to the surface. It is impossible to tell into what subterranean cavern the

deep undertow and currents take them and then the waters are filled with sharks which swim in both depths and shallows." However, according to the United States Coast Guard on April 20, 1993, the primary reason that corpses in Alaskan waters are not recovered is that "cold water doesn't allow them to decompose. Decomposition creates gases which bloat bodies and make[s] them buoyant enough to float to the surface."

18. *Victoria Daily Colonist (Victoria, British Columbia)*, "Shipwreck and Death," February 22, 1898.

3. THE *SEATTLE TIMES* AND THE *POST-INTELLIGENCER* DO BATTLE

19. *New York Times*, "Klondike Steamer Lost," February 15, 1898.
20. *Seattle Times*, "A Tale Most Terrible," February 18, 1898.
21. *Seattle Times*, "A Most Cowardly Attack," March 1, 1898.
22. *Seattle Times*, "*NEVADA*," February 15, 1898.
23. *Seattle Times*, "Divers Find the Hole," February 25, 1898.
24. *Seattle Times*, "Pathetic and Startling," March 10, 1898.
25. *Seattle Times*, "*Clara Nevada* Rumor," February 15, 1898.
26. *Seattle Times*, "About the *Clara Nevada*," April 5, 1898, and April 7, 1898.
27. *Scientific American*, October 15, 1898, 251. This number is suspect; however, some sources show as many as sixty boats operating on the Yukon River in 1898.
28. *Post-Intelligencer*, "Solving Mystery of *Clara Nevada*," March 13, 1908.
29. Gordon Newell to Steven C. Levi, personal correspondence, University of Alaska, Anchorage, archives.
30. James L. Nelson, conversations with the author, September 6, 1991 and May 26, 1992.

4. "CURIOUSER AND CURIOUSER"

31. *Daily Alaska Dispatch*, "Relics from *Clara Nevada* Brought Here," June 28, 1916.
32. *Dyea Trail*, "The *Clara Nevada*," March 11, 1898; *Post-Intelligencer*, "Boilers did not Explode," March 3, 1898.
33. *Daily Alaska Dispatch*, "Skull Found May be from Burned Boat," May 7, 1918. In an earlier article—*Dyea Trail*, "Body Found," March 11, 1898—it had

been stated that the mother of George Foster Beck was "bearing the expense of the search" for Beck but not that there was a reward for recovery of the body. Of absolutely no historical significance but for a passing note of humor, the man who made the discovery of the skull was William Swimlyornam.

34. Legend has it that the *Seaolin* was supposed to have been named "*Sea Lion*" but the painter had visited a few saloons on his way to work. Boehm, *Glacier Bay*, 40.

5. THE SEARCH FOR ANSWERS

35. *Anchorage Times*, "Miner Killed at Denali Mine," May 31, 1935. This Paddy McDonald's real name was Patrick Joseph McDonald. He was born in April 1874. It is doubtful that this is the same Paddy MacDonald of Teller, for the *Teller News* referred to him as an "old timer." In 1903, this McDonald would have been twenty-nine years old, hardly an old-timer.

36. *Post-Intelligencer*, "*Rosalie* from the North," February 15, 1898.

37. The signature was found on the legal papers with regard to the *William H. Evans*.

6. WHAT HAPPENED TO C.H. LEWIS?

38. Shipping Articles in RG21, USDC-Alaska, Civil #859, *Alaska Packers v. Steamship William H. Evans* at the National Archives–Alaska Region. For no known reason, it is located in the Third District Court Civil Records, circa 1946.

39. Ibid.

40. Lewis Klondike Exploration Company incorporation papers, state of West Virginia. The firm was involuntarily dissolved in 1947 because no request was made to extend the life of the corporation.

41. Wreck report, Alaska Packers Association Records; National Archives to Steven C. Levi, correspondence, October 3, 1990, *Clara Nevada* files, Steven C. Levi papers, University of Alaska, Anchorage, archives.

42. *Post-Intelligencer*, "*Eugene* in Courts," May 25, 1898. The copy of the paper in which this story appeared was so faded that it was virtually unreadable.

43. *Post Intelligencer*, "On Two River Steamers," June 22, 1898.

44. *Post-Intelligencer*, "Twelve River Steamers," May 27, 1898; *Post-Intelligencer*, "Today's Alaska Fleet," June 20, 1898.

45. *Post-Intelligencer*, "Yesterday's Big Exodus," June 22, 1898.
46. *Post-Intelligencer*, "Strange Tale of the Seas," August 21, 1898.
47. Arthur E. Knutson, "*Evans*" (unpublished, Anchorage Museum of History and Art, 1990).
48. *Seattle Times*, "Yukon River Boats," August 31, 1898; *Seattle Times*, "At the Ocean's Bottom," July 20, 1898; *Seattle Times*, "Trouble with Tows," July 7, 1898.
49. *Post-Intelligencer*, "Strange Tale of the Seas," August 21, 1898.
50. Knutson, "*Evans*." His source of information is listed as the *Post-Intelligencer* for August 26, 1898, and September 17, 1898. I have no reason to doubt the veracity of these quotes, but I could not find articles in the paper on those days. However, on the microfilm I was researching, the paper was illegible. Assuming that Knutson read the stories from a bound volume, the stories would be easily identifiable. Knutson's other references to articles, allow me to add quickly, are accurate, though they may be a day or two off.
51. *Post-Intelligencer*, "Along the Yukon River," July 15, 1898. I do not know why the newspaper referred to the "low water line" rather than a high water mark. Equally mystifying is the article "Yukon River Falling," in the August 27, 1898 edition of the *Post-Intelligencer*.
52. *Baltimore City Directory*; Melinda K. Friend, assistant manuscripts librarian, Maryland Historical Society, to Steven C. Levi, personal correspondence, August 2, 1990, University of Alaska, Anchorage archives.
53. United States Bureau of the Census, 1900.
54. Advertisement in the *Daily Astorian*, October 15, 1880. I could not locate a "Nestucca Bar," but I was able to find a Nestucca River; I presume the bar is at the mouth of the waterway. The Nestucca River enters the Pacific Ocean at Pacific City and crosses Highway 101 between Hebo and Beaver.
55. *Post-Intelligencer*, "What Steamer Is This?" February 15, 1898; *Lewis & Dryden's*, 290; *San Francisco Call*, "A Horror at Sea," December 22, 1887; Sacramento *Daily Record*, "Awful Disaster," December 22, 1887.
56. *Lewis & Dryden's*, 254, 255, 333; *San Francisco Call*, "The Steamer *Duncan*," September 21, 1885; *San Francisco Call*, "The Wrecked *Alexander Duncan*," September 11, 1885; *San Francisco Call*, "The Steamer *Alexander Duncan* Saved," September 20, 1885.
57. *San Francisco Alta*, "Wreck of the Steamer *Salinas*," November 5, 1871; *San Francisco Alta*, December 19, 1871; *San Francisco Alta*, December 20, 1871.

58. *Lewis & Dryden's*, 268. However, a check of the articles of incorporation of the Oregon Railroad and Navigation Company, filed on July 16, 1896, reveals three names on the documents: Henry Failing, E. McNeill and I.W.W. Cotton.
59. Bob De Armond to Steven C. Levi, correspondence, April 21, 1989, University of Alaska, Anchorage archives.
60. *Alaskan (Sitka)*, "The *Idaho*'s Officers," September 29, 1888.
61. Gibbs, *Shipwrecks of Juan de Fuca*, 158. *Lewis & Dryden's*, on page 356, mentions C.H. Lewis being the captain of the *Michigan*, but it is not clearly stated that he was indeed captain when the vessel made its "wild run" down the coast. "Caught on Fire Out at Sea," an article in the November 2, 1890 edition of the *Daily Astorian*, states that the captain was S.F. Graves. There are other articles on the *Michigan* on October 20 and November 3 in the *Daily Astorian*.

8. Mass Murder and Robbery or an Accident?

62. *Post-Intelligencer*, "Foster Beck's Body Found," a March 12, 1898 dispatch from Manaimo, British Columbia, that was published the next Sunday.
63. *Alaska Miner*, "Discovery of Purser Beck's Body on Chilkat Island," March 12, 1898.
64. *Seattle Times*, "*Clara Nevada* Rumor," February 15, 1898.
65. Record Group 104, Seattle Assay Office, Register of Deposits, April 2, 1900, to December 29, 1900, Seattle Assay Office Records, National Archives, Pacific Northwest Branch, Seattle, Washington.
66. *Seattle Times*, "Was on the *Clara Nevada*," March 17, 1898. As a tidbit of unusual information, according to Boyce, the stern lifeboat, which Boyce called the "rotten skiff which would not have floated to be rowed," was "PASSED BY THE INSPECTORS AS A LIFEBOAT." Those are Boyce's all caps, not the author's. I find this hard to believe since the *Clara Nevada* had been ordered by the inspection service to add two new lifeboats before it left and got rid of the grass preservers.
67. *Post-Intelligencer*, "None to Tell the Story," February 18, 1898.
68. *Seattle Times*, "Another Phase of It," February 23, 1898.
69. *Dyea Trail*, "An Echo from the *Clara Nevada*," May 7, 1898; *Alaska Miner*, "An Echo from the *Clara Nevada*," April 23, 1898. I was unable to find any information on J.H. Burns. Should it turn out that Burns specialized in strong, durable, expensive, one-of-a-kind lifeboats, I feel that will strengthen the case for first-degree murder.

70. Deputy Collector C.S. Hannum to Collector of Customs, Sitka, letter, February 18, 1898, Alaska State Library, Juneau.
71. *Dyea Trail*, "The *Clara Nevada*," March 11, 1898.
72. *Post-Intelligencer*, "Foster Beck's Body Found," March 13, 1898.
73. Wreck Report, Alaska Packers Association, microfiche, Z.J. Loussac Library, Anchorage.
74. The business archives in both Washington and Oregon are not indexed in a way that it would be easy to search for a specific name. My search of those records was by mail and by taking a "pot shot" at what names I believe the companies might have used, such company names included "Klondike," "Lewis," "MacDonald," "Rogers" and "Skagway." There could very well be a corporation started in 1898 with $165,000 with familiar names on the board of directors. I, however, was unable to find that documentation.
75. *Post-Intelligencer*, "The Discharged Gold Cargo of Treasure Ship *Roanoke*," July 19, 1899.

APPENDIX

76. *Seattle Times*, "Still Another Victim," March 19, 1898. Letters relating to Banks were published on the Internet in 1999. Copies are available at the archives of the Anchorage Museum of History and Art.
77. This inclusion could be a mistake, as a "William Maloy," another corpse, was also being transported south by the *Rosalie*. *Seattle Times*, "*Rosalie* in Port," February 15, 1898.
78. *Seattle Weekly Times*, "Pathetic and Startling," March 10, 1898. This article was copied by a researcher in Seattle and the identification line indicates "*Seattle Weekly Times*," but I suspect it is the *Seattle Daily Times.*
79. The *Victoria Daily Colonist*, in a story from the *Seattle Times*, lists a number of crew that was not included because of the vagueness. For the record, these included the following "four quatermasters," "two sailors," "three firemen," "three coal passers," "one night watchman," "four other cabin boys," "two kitchen helpers" and "three Chinese cooks." The story in the *Victoria Daily Colonist* appeared on February 20, 1898.
80. This could have been "Fred" Emery. According to the *Seattle Times*, Fred Emery had left the *Clara Nevada* in Dyea to help a Boston party build a hotel. Fred's letter to his father was, in the most classical way, a godsend.
81. *Seattle Times*, "Was on the *Clara Nevada*," March 17, 1898.

82. *Post-Intelligencer*, "Well Known Men on the Lost Vessel," February 17, 1898. The actual quote was, "It is possible the freight clerk, George Rogers, may have remained in Skagway and it is also learned that several members of the crew intended to desert in Alaska."

Bibliography

Secondary Documentation

Austin, Basil. *The Diary of a Ninety-Eighter.* Mount Pleasant, MI: John Cummings, 2007.

Becker, Ethel Anderson. *Klondike '98.* Portland, OR: Binfords & Mort, 1972.

Berton, Pierre. *Klondike: The Last Great Rush.* Toronto, ON: McClelland and Stewart, 1972.

Boehm, William D. *Glacier Bay.* Anchorage, Alaska: Northwest Publishing Company, 1975.

Buranelli, Vincent. *Gold: An Illustrated History.* Maplewood, NJ: Hammond, Inc., 1979.

Cantwell, First Lieutenant R.C.S. *J.C. Report of the Operations of the U.S. Revenue Steamer Nunivak.* Washington, D.C.: Government Printing Office, 1902.

Curtin, Walter R. *Yukon Passage.* Caldwell, ID: Caxton Printers, 1938.

"Dangerous Passage: Gold Rush Shipwrecks of 1898," *Alaska Historical Commission Studies in History* no. 194 (June 1986).

Downs, Art. *Paddlewheels on the Frontier.* Seattle, WA: Superior Publishing, 1972.

Gibbs, James A. *Pacific Graveyard.* Portland, OR: Binfords & Mort, 1950.

———. *Shipwrecks of Juan de Fuca.* Portland, OR: Binford & Mort, 1968.

Hunt, William R. *Distant Justice: Policing the Alaska Frontier.* Norman: University of Oklahoma Press, 1987.

———. *North of 53: The Wild Days of the Alaska-Yukon Mining Frontier, 1870–1914.* New York: Macmillan, 1974.

Johnson, James Albert. *Carmack of the Klondike.* Fairbanks, AK: Epicenter Press, 1990.

Lewis & Dryden's Marine History of the Pacific Northwest. Seattle, WA: Superior Publishing, 1966.

McCurdy, H.W. *The H.W. McCurdy Marine History of the Pacific Northwest.* Seattle, WA: Superior Publishing, 1966.

Newell, Gordon. *SOS North Pacific.* Portland, OR: Binfords & Mort, 1955.

Orth, Donald J. *Dictionary of Alaska Name Places.* Geological Survey Professional Paper 567. Washington, D.C.: Government Printing Office, 1967.

Place, Marian T. *The Yukon.* New York: Ives Washburn, New York, 1967.

Rogers, Alfred. *Shipwrecks of British Columbia.* Vancouver, BC: Douglas & McIntyre, Vancouver, 1973.

Scott, R. Bruce. *"Breakers Ahead!"* British Columbia: Review Publishing House, 1970.

Stummer, Harold Merritt. *This Was Klondike Fever.* Seattle, WA: Superior Publishing, 1978.

Wells, R.E. *A Guide to Shipwrecks along the West Coast.* Sooke, BC: Self-published, 1981.

Winslow, Kathryn. *Big Pan-Out: The Story of the Klondike Gold Rush.* New York: W.W. Norton, 1951.

PRIMARY DOCUMENTATION

Alaska Packers Association Records. Microfiche, Z.J. Loussac Library, Anchorage, Alaska.

Baltimore City Directory. Detroit, MI: R.L. Polk & Company, 1888.

Deputy Collector (Revenue Cutter Service?) letter to "Collector of Customs" in Sitka, February 18, 1898. Alaska State Library, Juneau.

Judicial records of Alaska. State of Alaska Archives in Juneau.

Lautaret, Ronald L. "The Juneau, Alaska Board of the Steamboat Inspection Service, 1898–1910." Master of arts thesis, Western New Mexico University, April 1979.

L.H. Bayer's References from Alaska Newspapers in the possession of the Alaska State Library, Juneau.

Levi, Steven C. Personal correspondence, 1980–2010. University of Alaska, Anchorage.

Niendorff, Hazel W. Scrapbook collection of the *Clara Nevada*, Museum of History and Industry, Seattle, Washington.

Ship Registries and Enrollments, Port of Portland, Oregon, 1869–1941. Oregon Historical Archives, Portland.

United States Government. Customs Service. Original documentation from St. Michael, 1898–1905. National Archives Alaska District, Anchorage.

———. Dispatches of the U.S. consuls in Dawson City. Federal Records Center, Anchorage, Alaska.

———. Index by district to U.S. Coast Guard Reports of Assistance. National Archives Alaska District, Anchorage.

———. Records of Alaska customhouses. National Archives Alaska District, Anchorage.

———. Records of entries and liquidations, Eagle, Alaska. National Archives Alaska District, Anchorage.

———. Revenue Cutter Service. Alaska File. National Archives Alaska District, Anchorage.

———. Treasury. "Report upon the Customs District, Public Service and Resources of Alaska Territory by William Gouverneur Morris, 1879." Z.J. Loussac Library, Anchorage, Alaska.

———. U.S. Coast Guard casualty and wreck reports. National Archives Alaska District, Anchorage.

NEWSPAPER AND MAGAZINES

Alaska Forum

Alaska Journal

Alaska Magazine

Alaska Miner (Juneau)

Alaska Mining Record

Alaska Searchlight (Juneau)

Alaska Sportsman

Alaskan (Sitka)

Daily Alaska

Daily Alaskan Dispatch

Dawson Daily News

Dyea Press

Dyea Trail

Fort Wrangell News

New York Times
Nome Nugget
Olympia Washington Standard
Oregonian
Petersburg Weekly
Pony Express
Rampart Miner
Seattle Daily Times
Seattle Post-Intelligencer
Seattle Weekly Times
Skagway News
Stikine River Journal
Tanana Directory
Teller News
Victoria Daily Colonist (Victoria, British Columbia)
White Horse Star
Yukon Press
Yukon Sun

About the Author

Steven C. Levi is a historian and freelance writer based in Anchorage, Alaska. Levi has lived in Alaska for thirty-three years and has more than thirty books in print, ranging from Alaska history to Westerns, how-to and self-help to poetry and aviation to biography. Levi has also developed educational software, written screenplays and developed creative media presentations using poetry and history, a sample of the last being on his website, parsnackle.com (See the "Phantom Dogsled"). Levi's scholarly articles have appeared in such publications as *Journal of the West*, *Pacific Historian*, *East Texas Historical Quarterly*, *Western Folklore*, *Southern California Quarterly*, *California* and *Labor History*.